S0-AXR-234

PRAISE FOR THE PREVIOUS EDITION...

I thought I knew a lot about my iPhone XS until I read Rich's book. This book is filled with tips to help get the most out of this $1100 phone.

- LONESTAR10 ON AMAZON

As an I phone user of many years I knew that there are many things my phone was capable of but did not know how to access. this book is well written and easy to understand, very helpful

- FRANK ON AMAZON

I keep going back and re-reading to find yet another helpful feature or semi-hidden tool on the iPhone. Very helpful, useful volume. No need to read straight through, just jump around and read a tip or two or three a day.

- J BACON ON AMAZON

101 IPHONE TIPS & TRICKS

UNLOCK THE USEFUL, TIME SAVING AND FUN FEATURES IN IOS 13

RICH DEMURO

Copyright © 2019 Rich DeMuro, LLC

All rights reserved.

No part of this book may be reproduced in any form or by any electronic or mechanical means, including information storage and retrieval systems, without written permission from the author, except for the use of brief quotations in a book review.

All product names, trademarks and registered trademarks are property of their respective owners. All company, product and service names used in this website are for identification purposes only. Use of these names, trademarks and brands does not imply endorsement.

Apple and iPhone are trademarks of Apple Inc.

richontech.tv

Dedicated to those who believe
that anything is possible.

"You should write a book on this stuff"

- EVERYONE I MEET

CONTENTS

EVERYTHING ELSE

INTRODUCTION

When I set out to write the predecessor to this book, a paperback edition of *101 Handy Tech Tips for the iPhone,* I really had no idea if anyone would ever read it. I went home after my mornings at KTLA, turned down the lights in my office, and wrote. And wrote. Then I edited.

What started as a single tip - how to activate the iPhone's built in magnifying glass feature - turned into a book. The more I explored all of the fun little features of the iPhone, the more I wrote.

It became a treasure hunt. I felt that people use about 1% of the features on their phone and I wanted to change that.

I submitted the book on Amazon just before Black Friday 2018 and waited.

Next thing you know, the orders starting coming in. At one point, *101 Handy Tech Tips for the iPhone* was the 19th bestselling book on Amazon.

Sure, they do those list by the hour, but clearly, I had touched a nerve.

But the thing I braced for most was the feedback. Sure, you can sell a book, but I most curious how people would respond to their purchase. In today's age of Twitter, Yelp, Facebook and Amazon reviews, I knew that the feedback wouldn't be far behind.

The reviews were overwhelmingly positive - both on Amazon and through email, social media and more. Many people got in touch with me personally to let me know how the book clued them in on useful iPhone features they never knew existed.

What's in your hands right now is a culmination of the experience of writing the first paperback. Some of you might remember that I originally published the book as an ebook. That was good, but the number one request from readers was to provide a printed copy - something you could thumb through with your phone nearby.

This paperback contains tips pertaining to the latest version of the software running on the iPhone - iOS 13 - along with tips I've deemed as "classic" from the previous editions.

I set out to write a followup that would highlight the new features of the operating system but also embrace the things I think everyone should know about this amazing little gadget.

Now, let's get to it, shall we?

HOW TO USE THIS BOOK

This is not necessarily a user's guide, as it does not explain every feature available on the iPhone. Also, it assumes a basic working knowledge of the phone: how to turn it on, what an app is, how to make a phone call, send a text and take a picture.

But pretty much everyone knows how to do those things on the iPhone.

You're here because you want to take it a step further. You want to get more out of this device that you paid a lot of money for.

The tips inside this book are organized first by the new features in iOS 13 and then the rest of the stuff I think you should know.

There are a lot of iPhone models. The tips in this book take advantage of the latest software update as of this writing, which is iOS 13. If you haven't updated your phone, some of the features might not work.

Also, the iPhone is now "forked" into two types: those with a home button and those without a home button. The main difference is for bringing up Control Center. On an iPhone with a home button, you swipe up from the bottom of the screen to accomplish this. On an iPhone without a home button (iPhone X and above), you swipe down from the upper right hand corner to accomplish this.

I've tried to make doing the tips as easy as possible. I use this formatting to signify the various steps:

Settings > **General** > **Date & Time** > **Set Automatically**

This means you would start by tapping the **Settings** app, then tap the section labeled **General**, then tap where it says **Date & Time**, then look at the toggle for **Set Automatically**.

I also try to keep the formatting the same as the iPhone. If a button or menu is written a certain way, I'll use the same language so you can easily identify it. If something is in all caps, I'll write it in all caps.

Some buttons are tough to describe, like "three little lines" or "a circle with an i in the middle." I'll make it as easy as possible for you to identify what I'm talking about.

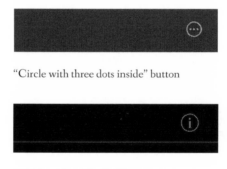

"Circle with three dots inside" button

"Circle with an i inside" button

My goal in all of this: to help you unlock more of the functionality of your phone and to make you feel confident using it.

The more you know about the iPhone's capabilities, the more you can make it your own. Also, you'll be less frustrated when something doesn't seem to be going your way.

I hope you enjoy reading this book as much as I enjoyed creating it.

WHAT'S NEW IN IOS 13

1 / STEP INTO DARK MODE

WHO WOULD HAVE THOUGHT that one of the most celebrated new features of iOS 13 - the one that got the most applause at Apple's Worldwide Developers Conference 2019 - would be something called "dark mode."

The name says it all. It makes everything on your phone screen appear dark.

I'm not sure where the fascination with dark mode started, but it probably has it roots in coding. Coders spend long hours at their computer screens typing in lines of programming, and staring at a bright screen isn't so easy on the eyes after a while.

Dark mode is also handy for the rest of us who are (unfortunately) checking our email, social media feeds and more right before bed.

Once activated, Dark Mode will change the way your phone screen looks to, well, dark.

In iOS 13, most of the apps that Apple built will appear dark when the mode is activated, but when it comes to other apps, they may or may not appear dark depending on how they're coded.

Expect big name apps to adapt to the new mode first, followed by the smaller ones.

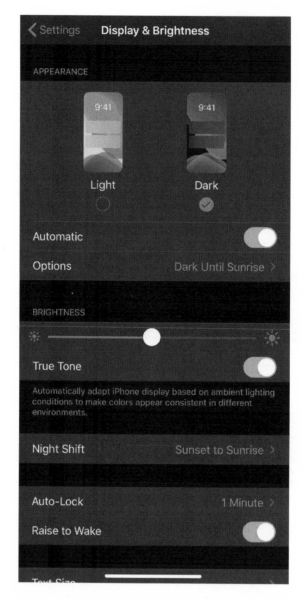

Activate Dark Mode to change the look and feel of your screen

To activate Dark Mode manually, go into **Settings** > **Display & Brightness**.

Under the **APPEARANCE** section, tap the option for **Dark**.

Instantly, your screen will look much darker.

Now you have a decision to make: do you want your phone to remain in dark mode all day long, or would you rather it follow a set schedule, like sunset to sunrise? You can even create your own schedule based on what works for you.

If you want your phone to stick exclusively to Light or Dark mode, select the mode you want and then be sure the **Automatic** toggle is off.

If you want your phone to switch between Light and Dark modes depending on the sun, turn the **Automatic** toggle on.

Under **Options**, you can customize whether your mode follows Sunrise and Sunset, or a schedule you choose.

Remember, not every app you open will automatically have a dark mode. Developers must update their app to work with this feature. This means - especially in the early days of the feature - not all of your favorite apps will be dark.

Slowly, but surely, they'll catch on and follow the same schedule you set in your iPhone settings.

If you're a vampire, bat or otherwise don't like a lot of light, life just got a whole lot easier.

2 / PROTECT YOUR PRIVACY WITH NEW LOCATION ACCESS OPTIONS

EVERY APP WANTS access to your location at all times. The reason? They can use this information to build a more precise profile of you and have better luck at selling you stuff.

Of course, there are many benefits to sharing your location with an app - it's how a car sharing service finds your house or a weather app shows your local forecast.

Apple understands that location access can be a double edged sword, which is why they're giving users more precise sharing options in iOS 13.

For the first time, you can grant access to your location just once when you open an app. This means that even if you download an evil flashlight app that wants access to your GPS location, they won't get very far with it. Pick the new option to **Allow Once** and as soon as you close out the app, their GPS stream is cut off.

This can be handy in a variety of situations where you don't feel that it's necessary to share your location with an app 24/7.

There are now up to 3 location sharing options available to you when you first open an app.

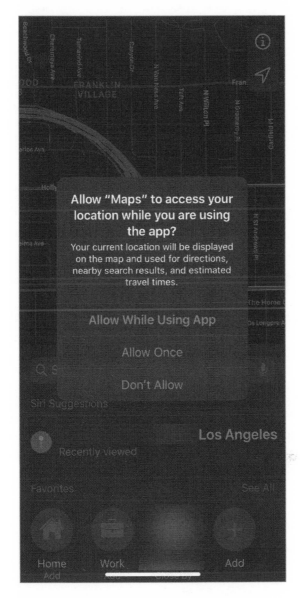

You have options when it comes to sharing your location with apps

The first is **Allow While Using App**. This means that as long as you are using the app, it has access to your location. This can be handy for a ride sharing service or a mapping app.

The next, new option is **Allow Once**. This means that the app will get your current location this time, but that's it. If it wants to know where you are in the future, you will have to grant it access once again.

The final option is **Don't Allow**. Choose this and an app does not get access to your location.

Keep in mind, once you make a selection you can always change your mind later.

Just go into **Settings > Privacy > Location Services** and you'll see a list of your installed apps. Tap one to change it's location access setting.

Another new thing you'll notice in iOS 13 is that your phone will alert you every so often after an app accesses your location several times.

You'll get a pop up message along with a map revealing all of the places that particular app has accessed your location in the background. There's a choice to continue to allow this, or restrict location access to only while using the app.

These pop ups are a helpful reminder of how apps build a better database of who we are and where we go. Read them carefully, and choose the best privacy option based on the functionality of that app.

For instance, it makes sense for the weather app I prefer to always have access to my location so it can tell me when it's about to rain where I am. It doesn't make sense for the salad place I order from occasionally to always know where I am, even when I'm not actively placing an order.

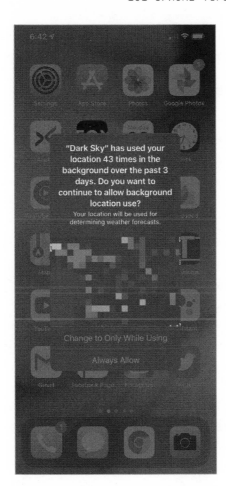

These pop-ups look scary, but read them carefully and make a decision that's best for you

3 / CREATE A NAME AND PROFILE PHOTO

iMESSAGE MASQUERADES as text messaging but it's really a social network of its own. Just think about how many people use it on a daily basis.

Now it's getting a feature that will help identify you to the other people you message. You can now set a profile photo (or initials) along with a name so people instantly know it's you.

To set it up, you have to be signed into iCloud and iMessage must be turned on. Chances are, that's probably the way your phone is already set up.

Once that's sorted out, open the **Messages** app - you know, the one with the green bubble. Ironic, since Apple reserves green bubbles for, ahem, Android users. You would think they would have changed the app's icon to blue by now.

Next, hit the button with three dots next to the button to compose a new message.

Messages

Tap the option to **Edit Name and Photo** and you can begin the process of personalizing how you'll appear to other people you message in iMessage.

You'll probably want to start by entering in your name. It seems as if you can write anything in here you want, but use your discretion. You might message your boss from this account at one point, and while Princess Peach might be appropriate for your friends, it might be embarrassing to explain your nickname to others.

Above your name you can tap to edit your photo. Your iPhone might have some suggestions for you, including any recent Animoji, Memoji, selfies and initials.

If nothing is striking your fancy, there are still several more options you can go with.

If you tap the **Camera** icon under **Suggestions**, you can snap a photo or selfie of yourself. You even get an option to choose a filter before you finalize your photo.

Another option is to tap **All Photos**. This will let you choose a picture from your camera roll.

When you finally decide on your picture, Apple will show a message box letting you know that your Apple ID and your contact card will now display this image.

To share your newly personalized name and photo with the people you iMessage, be sure to toggle **Name and Photo Sharing** on.

By default, your name and photo will automatically be shared with your contacts, but you also have the option to have your iPhone ask you every time you send a message if you want to share your personalized info.

The next time you text someone, they'll be able to see your name and profile picture if you allow it. This means no more signing texts with your name or getting those awkward "who is this" responses back.

Once your Name and Photo is setup, if you change your mind,

you can always revisit these settings either from within the Messages app or by heading into **Settings** > **Messages** and tapping the option for **Share Name and Photo**.

4 / SWIPE TO TYPE

THE LATEST FEATURE TO hit the Apple Keyboard isn't necessarily new, but I'm already in love with it. It's the ability to swipe to type!

Sure, a company named Swype introduced it for Android way back when, but who's keeping score? And yes, there are already many keyboards in the App Store that offer the feature, but this time, it's baked right into the keyboard you probably use most - the one built into your iPhone.

The feature should be turned on by default when you install iOS 13. To try it out, just bring up the keyboard in a place where you would normally type. It could be an email, a text, Facebook post or whatever.

Now, instead of tapping the keys, just trace the word you want to spell without lifting your finger off of the keyboard.

When you are finished, lift your finger and you should see the word you just traced appear on the screen. If not, there will be a few close suggestions on top of your keyboard.

It might take some getting used to, but this is a fun, fast and efficient way to type. The best part - you can do it one handed, while you walk! Not that I would condone that sort of behavior.

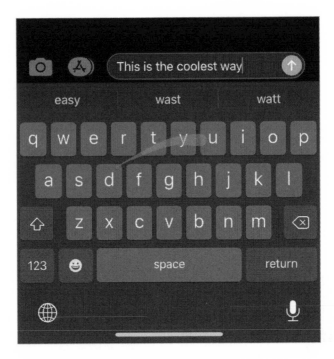

Just swipe your fingers over the keys to spell out words

Don't fancy this new style of typing? You can turn it off by going into **Settings** > **General** > **Keyboard** and flipping the switch next to **Slide to Type**.

Before you do, just remember, swiping and typing are not mutually exclusive. You can type out a word the old fashioned way, then you can swipe to type the next word. Heck, you can even use Siri to dictate the next word. It's up to you!

Personally, I find swiping to be a really fast way to type. I'm always amazed at how accurate the predictions are based on such loose tracing.

5 / ACCESS THE EMOJI KEYBOARD
FASTER

FEELING HAPPY, sad or crazy? You can express yourself in Emoji faster than ever in iOS 13. That's because there is now a dedicated button on the keyboard that will take you right to the good stuff.

Previously, to get to your Emoji keyboard, you had to press and hold a world button, then tap Emoji to get to the fun little icons.

Now, all you have to do is press the **Smiley face button**, which is always on display in the Apple Keyboard.

If you have multiple keyboards enabled, the button will appear to the left of the space bar.

If you just have the Apple Keyboard enabled, it will appear under the spacebar.

Either way, all you have to do is tap it once and you're there!

The only question is, how fast can you build an entire sentence using just emoji?

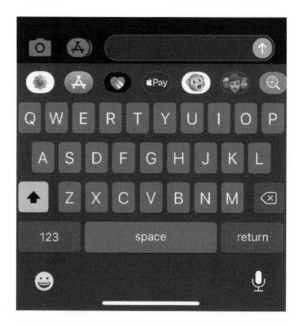

Tap the smiley face to access your emojis fast

6 / PICK UP THE CURSOR TO MOVE IT

RIGHT BEFORE THE last version of this book was about to go on sale, there was an iPhone tip that went viral. It involves pressing and holding the space bar down on the keyboard, then moving the cursor around the screen like a trackpad mouse.

Now, there's an even cooler way to move the cursor to where you need it on the screen: just pick it up and drop it down whenever you need it.

To try it, open up something like a note in your Notes app with a lot of text in it. Or just start typing out an email.

Either way, locate the blinking cursor on the screen. Now, place your finger on it and without lifting your finger off the screen, drag the cursor around.

You can virtually drag and drop the cursor wherever it needs to go.

Even better, it will automatically snap to words and lines if you want it to. Just get it "close enough" the beginning or end of a word or some open space after a sentence, and it will position itself just where your brain wants it, like magic.

Need to move the cursor? Just tap and drag it to a new
location

This trick works in nearly all of the places I've tried it on the
iPhone. If there's a cursor blinking on the screen, chances are you can
just drag and drop it to a new position instantly.

THE FIRST TIME you go to rearrange your apps in iOS 13, you might notice something kind of strange: the feature doesn't work like it used to.

For as long as I can remember on the iPhone, you would press and hold on an app and they would all start to wobble, indicating that you could now pick them up and move them around into different positions on the screen.

Try the same thing now and the apps no longer wobble instantly. You're now presented with several options, including a new one to **Edit Home Screen**.

Tap it and *then* you will see the familiar wobble.

Now, you're free to move apps around or hit the **x** to delete one from your phone altogether.

You an also easily delete an app by long-pressing its icon and choosing the **Delete App** option.

By the way, rearranging apps the "old fashioned way" still works in iOS 13, it's just a little trickier to access.

Here's how to do it: in one movement, press hard on an app until you feel a slight vibration confirmation. Then, without lifting your finger off the screen, drag the app a bit.

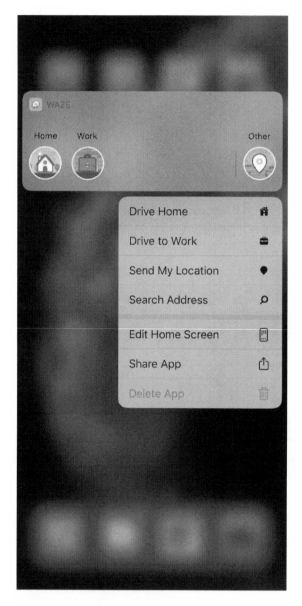

Long press on any app icon to see the option to edit your
home screen and quickly delete apps

You'll notice that your screen instantly starts to wiggle and you
can move around all of your apps.

IF YOU JUST TRIED THE last tip, then you probably got a little preview of this one. What started out as Force Touch and then became 3D Touch is now Haptic Touch in iOS 1 3.

C'mon, make up your mind, Apple!

The functionality has somewhat evolved over the years. What used to be a "hard push" is now more of a "tap and hold." Another way to think of it is a "long press."

Try it with an app icon. Instead of tapping it once to open up the app, tap and hold on the icon.

Suddenly, you'll get a new menu with shortcuts related to that app.

For instance, if you Haptic Touch the Camera icon, you'll get options to **Take Selfie**, **Record Video**, **Take Portrait** and **Take Portrait Selfie**. There's also the feature we talked about in the last tip, **Rearrange Apps**.

Haptic Touch on the Phone icon and you'll see options for **Create New Contact**, **Search For Contact** and **View Most Recent Call**.

This is a handy feature that I'm betting 98% of iPhone users don't actually use. Still, it can be a real time saver.

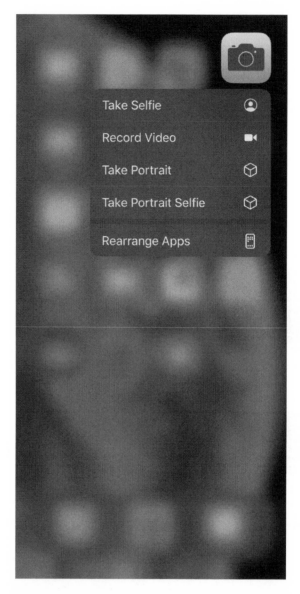

Long press on an app icon to see shortcuts related to that
app

For instance, if you wanted to create a new contact the standard
way, you would have to tap the Phone icon, then tap the Contacts
section, then tap the plus sign and start typing in the information.

With Haptic Touch, just press and hold on the Phone icon, then tap **Create New Contact**. It saves at least two taps, which doesn't sound like a lot, but don't forget about the time it takes you to locate all of the places you need to tap on the screen to complete the task.

You'll notice that some apps have more shortcuts than others, others have none at all.

Take a few minutes to press and hold your various app icons to see what kind of shortcuts they offer. Once you commit a few to memory, you'll find that everyday tasks are quicker if you use them.

Some of my favorites are **Take Selfie** on **Camera**, **New Note** in **Notes**, **New Message** in **Messages** and **Start Stopwatch** in **Clock**.

THIS COULD BE the single most useful feature built into iOS 13: the ability to send unknown callers straight to voicemail.

That's right - you might never have to answer a robocall ever again.

Now, before you think that this heavy handed approach to unknown callers might be too restrictive, let me explain how it works.

Apple is applying a level of artificial intelligence, or Siri smarts, to the call filtering. Basically, calls from unknown numbers will be silenced and sent to voicemail, but only if there is no evidence of that number anywhere on your phone.

To further explain: If a number calls that isn't in your address book, Siri will look for that number in your emails, your messages and recent outgoing calls. This means if you are expecting a call from a plumber and you've emailed with them in the past and their phone number appears in their email signature or in a text they previously sent you, your phone will still ring.

Another example: let's say you call a handyman from your iPhone and leave a voicemail. When that same phone number calls you back, your phone will still ring since your phone sees that your dialed that number recently.

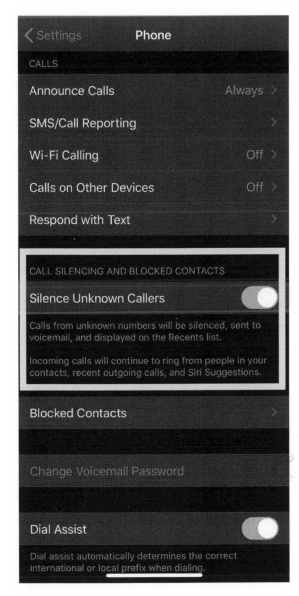

Turn on the option to "Silence Unknown Callers" to cut down on robocalls

If your phone can't find any evidence of the incoming number, the call will be sent straight to voicemail and you will get a Missed call notification. The call will also show up in your list of Recents.

I feel like this is a good solution if you're getting a lot of robocalls. Chances are, a majority of the people calling you are in your address book, their number is contained in an email you've exchanged or you dialed them at some point.

This setup would probably not work if you're in a business that gets a lot of cold calls. For instance: you're a handyman and people are calling you off Yelp. You probably don't want to turn this feature on.

The big wild card here is for parents. Let's say you have this feature turned on and someone from your kids school needs to get in touch with you for an emergency. If the call is coming from a teacher's cell phone or an unknown number that's not contained in any of your emails or messages, it would be sent straight to voicemail.

I'm just laying out all the possibilities here so you can make the best choice for your situation.

Clearly, this new feature is a great addition to the iPhone and one I suspect many will take advantage of.

To turn on Call Silencing, go into **Settings** > **Phone** > **Silence Unknown Callers**. Toggle the switch on to have calls from unknown numbers silenced and sent to voicemail. The call will still show up in your Recents list.

For a lot of people, this level of filtering will make sense and help cut down on a majority of calls offering everything from lower interest rates to extended vehicle warranties.

10 / CHOOSE WIFI & BLUETOOTH
     CONNECTIONS FAST

You're probably familiar with the routine to change or choose a WiFi network on your iPhone. You open Settings, scroll to the WiFi section, tap to see a list of available networks and tap to choose one.

Now, the process is getting a convenient shortcut. You can now choose your preferred WiFi network directly from Control Center.

Previously, your only WiFi option in Control Center was to turn WiFi off or on, but now you can go a step further and select which network you'd like to be on. It's faster than ever before.

To select a new WiFi network fast, just **go into Control Center** by swiping down from the upper right hand corner of your screen if you have an iPhone without a home button. If your iPhone has a home button, swipe up from the bottom of your screen to enter Control Center.

From here, press and hold on the little box that shows your connection icons - the one with the airplane, cellular signal, WiFi and Bluetooth symbols.

This will make the box larger and add labels to the controls.

Now, press and hold the icon labeled **Wi-Fi**. This will bring up a list of nearby Wi-Fi networks you can connect to. Tap a network to start the connection process.

Long press the WiFi button in Control Center to choose
a network instantly

Alternatively, you can choose the option for **Wi-Fi Settings...**
at the bottom of the box to take you to the Wi-Fi section of the
Settings screen you're familiar with.

The same process works for Bluetooth connections.

From the main Control Center screen, you can now press and
hold the **Bluetooth** icon to see all of your available Bluetooth
devices. Simply tap one to connect to it.

While we're here, let me explain the difference between turning
off WiFi and Bluetooth in Control Center vs the main Settings
screen.

When you tap to turn off Bluetooth or WiFi from Control
Center, that really just disconnects you from the current network.
Your iPhone may still connect to another known network when you
get near it, or even re-connect to the network you just disconnected
from if it's later in the day.

If you really want to turn WiFi or Bluetooth off and not have your phone automatically reconnect to a known network, you'll have to do it from **Settings** > **Wi-Fi** or **Settings** > **Bluetooth**.

11 / USE THE NEW ON-SCREEN VOLUME CONTROL

THE FIRST TIME you adjust the volume on your iPhone in iOS 13, you'll notice a completely different control. It's now located in the upper left hand corner of the screen and not in the middle.

While the old control had 16 levels of volume, the new one combines these 16 levels with a slider that also handles finer adjustments.

To try it out, press one of the **Volume keys** on the side of your phone to make the volume control appear on screen.

Each time you press a button up or down, the volume adjusts accordingly. Go ahead, count the 16 clicks.

After the first click, you'll notice that the volume slider minimizes to get out of the way. This way you can adjust the volume of your current Netflix binge without a giant volume icon covering the middle of the screen, like it used to.

But this is Apple, and they like to give you options, so there's another, less apparent way to control the volume.

Try clicking a **Volume button** again, but this time, slide your finger on the volume control. You can adjust all the way from mute to max.

The volume control gets a new location and look in iOS
13

This gives you finer control over the volume. And to really take it
to the next level (see what I did there?), press the volume key on the
side of the phone, place your finger on the volume slider, then

without lifting your finger, slide it around somewhere else on the screen.

You can now move the volume up and down without actually being on the virtual volume slider.

Sounds like Apple had way too much fun with this one.

See what I did there again?

Ok, I'll mute myself for now.

12 / MAKE YOUR MEMOJI LOOK JUST LIKE YOU

In iOS 13, you can now customize your Memoji in more ways than ever before. While previously you could pick a skin tone, hairstyle and eye color - the new and improved Memoji maker lets you customize every aspect of your look.

Have an eyebrow piercing? Want eyeliner? Nose piercing? Missing tooth? You can now even have earrings or AirPods hanging out of your ears.

To build your super custom Memoji, open up the **Messages** app and start a new message or jump into a chat you already have going. Now, look for the row of icons above the keyboard and tap the one that looks like a **Monkey**.

In here, you'll see any Memoji you created in previous iOS versions. Once you've settled on one, look for the button with the three little dots in it in the lower left hand corner. Tap it to bring up the option to **Edit** your Memoji. If you're worried you're going to mess up your previous masterpiece, hit the **Duplicate** button and you can start with a copy.

If you haven't created a Memoji yet, you can start from scratch by hitting the big **Plus Sign** button that says **New Memoji** under it.

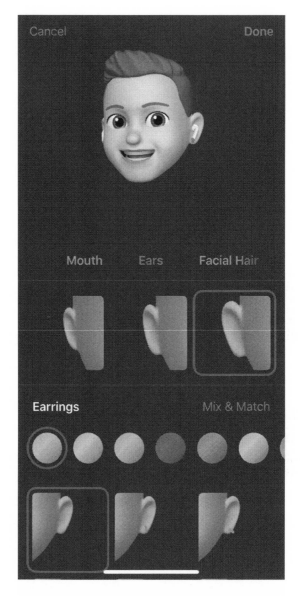

There are new options for creating your Memoji

Either way, you might want to carve out a little time to create your animated likeness. There are so many options, you might even want to grow a goatee just so you can represent in your little avatar.

When I scrolled through the Facial Hair section, I've never seen more types of beards in my life. There must be 100 variations. It could also be fun to create your alter ego, too.

When you're finished having a laugh and your face looks just right, hit the **Done** button and you're ready to send your Memoji out into the world - or at least in as many messages as you deem fit.

IF YOU'RE familiar with the concept of a Bitmoji, you'll understand Apple's latest feature that lets you turn your Memoji and their Animoji into "sticker packs."

I'm not going to go into my personal thoughts on the merits of using these sorts of things in your messages, but I'll just leave it at the fact that they are very popular.

In iOS 13, Apple automatically creates fun little expression filled stickers you can use throughout your apps - not just in messages.

To try it out, open the **Messages** app and compose a new message or reply to one you've already started. Look for the little ribbon of icons above the keyboard for a button that looks like three little Animoji faces.

Tap it and you'll see any Memoji you've already created, along with Apple's Animoji.

Now for the fun part: tap one of the little characters and suddenly you get a bunch of expression-filled "stickers" to choose from.

There is the crying face, the heart eyes, mind blown, sleeping, star eyes, single tear, kissy face, wink and more.

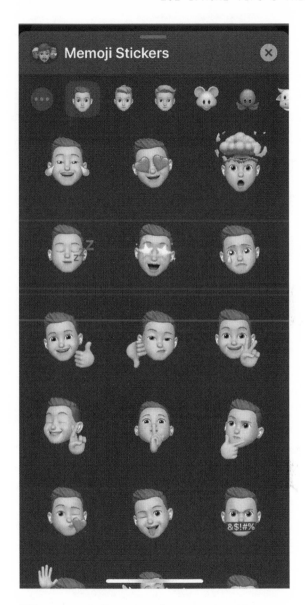

iOS 13 automatically generates fun, expressive stickers
based on your Memoji

Tap on any of them to insert this little slice of playfulness into
your message.

But wait, there's more!

While you're looking at the rows of options for any given character, swipe left or right to advance to the next character.

Even any Memoji you've created have a new range of expressions including Thumbs Up and Down, Peace, Fingers Crossed, Shhh, the shrug, facepalm and more. Apple really went out of their way to make these as fun as possible, and I bet you'll have a grand old time using them.

They're so cute, I might even have to change my stance on banning them from all of my messages.

THERE ARE two types of people out there: those who keep a highly organized address book and others who can't even identify half the people in there due to a mishmash of first names, initials and other half-entered data.

This new contacts feature will appeal to the organized group.

In previous versions of iOS, you could add a related name to a contact, including mother, father, son, daughter, child, partner, assistant, manager and a few others.

In iOS 13, Apple had a major brainstorming session regarding how people could be related to each other, because this feature has expanded to encompass hundreds of specific labels.

You can build an entire workgroup organizational flow chart here or create a family tree.

Not only do you get options like elder sister, youngest brother, and great-grandmother, there's also co-parent-in-law, great-grand-child or sibling's grandchild, grandnephew and many, many more. The best part is that each family member comes with a further descriptor so you can get highly specific.

You can get highly specific with your Labels in Contacts

Like, Aunt Jane (mother's brother's wife) or Uncle Jim (father's elder sister's husband). It's the kind of stuff that will come in handy at your next family reunion.

As if all the new options aren't enough, there is still an option to **Add Custom Label** or **other**.

To set the new relationships, open the **Contacts** app and choose a contact or create a new one. While you're editing, look for the field labeled **add related name**. Type in a related name or tap the **i** in the circle to choose an existing contact.

Then, tap where it says mother (or father, or parent, or something similar depending on how many fields you've added) and you'll see the list of labels to choose from. To really see them all, hit **All Labels** at the bottom of the list and have at it.

Who knows, maybe we'll see a "print family tree" option in the next version of iOS.

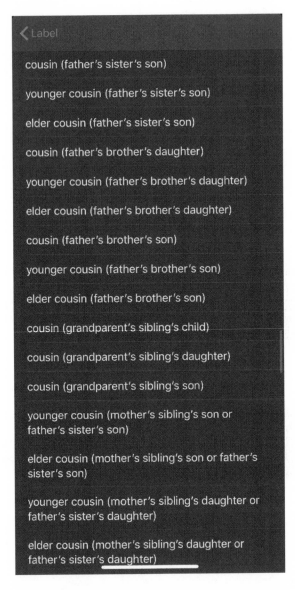

Told you these get real specific

15 / DOWNLOAD LARGE APPS ON A CELLULAR CONNECTION

IN THE PAST, if you tried to download a large app from the App Store and your only connection was cellular, you had a problem.

You had to wait until you had a WiFi connection to download the app if it was over a certain number of megabytes. Slowly but surely, Apple has increased this threshold.

In the previous version of iOS, that limit was 150 megabytes.

But, as apps get larger and cellular connections get more unlimited, Apple is now letting you download any size app you want over a cellular connection, as long as you allow it.

Apple's threshold for a large app in iOS 13 is now 200 megabytes.

The first time you try to download an app larger than this, you'll get a message explaining that the app is large and using data over a "cellular network may incur additional fees."

But, instead of forcing you to wait for a WiFi connection, you'll now get two new options: you can choose to wait for WiFi, or download right away using your cellular connection.

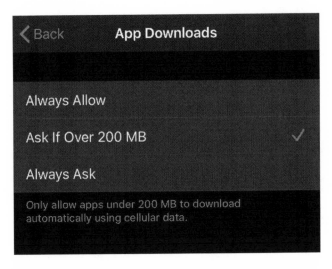

Download large apps, even on a cellular connection

It's nice to finally have a say in the matter. If you have a good data plan that can support the large download, you can use it. If not, you can take the box as a warning and save your data for something more useful.

Even if you choose one option or the other, you can always go back and change your mind.

To do this, head into **Settings** > **iTunes & App Store** and look for the section labeled **CELLULAR DATA**.

Tap **App Downloads** and you'll see all of your options: **Always Allow**, **Ask If Over 200 MB** and **Always Ask**.

Choose the one that works best with your data plan.

Keep in mind, this setting will apply to new apps you download as well as any app updates in the future.

At least now, the choice is yours!

16 / ASK SIRI TO TUNE INTO A
RADIO STATION

SIRI'S ABILITY TO tune in a radio station is getting a major upgrade in iOS 13. Before, she could summon the news from a few outlets and music from Beats 1, but after that, things got pretty limited.

Now, you can ask Siri to play just about any radio station you can think of and the live stream will begin instantly - no separate app downloads necessary.

Officially, Apple says there will be support for 100,000 radio stations from around the globe.

Stations from TuneIn, iHeartRadio and Radio.com are all represented. We'll just have to take their word on this one - I doubt anyone will try to tune them all in.

To try it, think of a radio station. Then, summon Siri and ask her to play it.

For instance, growing up in New Jersey, it was all about Z100. If I say, "**Hey Siri, play Z100 radio station**," the live audio stream starts playing automatically.

There's also a description under the station of where it's streaming from - in this case, iHeartRadio.

Similarly, I can say something like "**Hey Siri, play the radio station KFI**," and again, it starts streaming instantly.

You can ask Siri to play just about any radio station
instantly

Getting some stations to play might be trickier than others. In
LA, we have the popular station KROQ-FM. But if you ask Siri to

"**Play K-ROCK**," as it's pronounced, she might not get it right on the first time.

But if you say, "**play K-R-O-Q**" and pronounce each call letter individually, the station begins to stream from Radio.com, again no separate app necessary.

This is huge. The ability to call up this many radio stations without installing a single app is super convenient. This can be especially handy in the car when you're using CarPlay and don't feel like tapping the screen several times just to call up a streaming station.

It also puts terrestrial radio stations on a better playing field to compete with podcasts and other streaming music apps, which are also easy to call up using voice and on screen taps.

You've HEARD of Google Street View, now Apple has their answer with a new **Maps** feature called **Look Around**. It's similar to what Google offers, but built into Apple Maps.

It lets you take a look around the real streets surrounding a destination. If you want to confirm where a driveway is, exactly where a business is located or just get an idea of a neighborhood, you can use **Look Around** to take a virtual tour before you even get there.

Apple has done a lot of work on its Maps app. When it launched, it was a laughing stock. The company took the feedback to heart and has since sent special Apple Maps cars crisscrossing the nation to log streets and snap pictures. I've seen the cars in my neighborhood; perhaps you've spotted them, too.

To try the new **Look Around** feature, search for a city or a business. I'll use Zero Zero, my favorite pizza place in San Francisco. Seriously, I stop there every time I'm up there for a tech event.

If you open Apple Maps, type in Zero Zero and hit search, you should get the result. Now, look for the photo labeled **Look Around**. Tap it and you're instantly standing in front of the restaurant. You can use your finger to move around for a 360 degree view or double tap on the screen to move down the street.

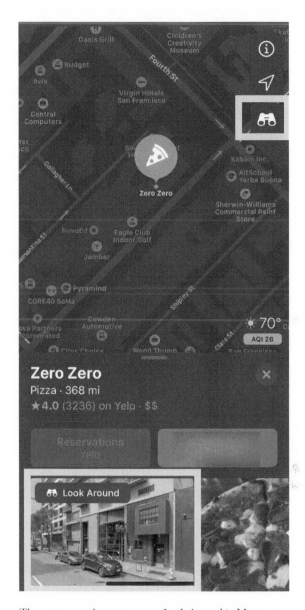

There are several ways to access Look Around in Maps

The further down the street you tap, the more of a virtual ride you can take.

You can explore other areas enabled with the Look Around feature by tapping the **two arrows pointed at each other** in the upper left hand corner of your screen while you're in Look Around mode.

This will take you back out to a map of the area. Now, move the map for a look around. Keep in mind that not every area has the Look Around feature enabled. You'll see streets in a city highlighted in blue if the feature is available there.

Look Around will appear in various places in Apple Maps.

You can search for a city from the Apple Maps main screen to see if it has Look Around enabled.

You can zoom into a specific street, and if Look Around is available for that area, you'll see an icon that looks like a pair of binoculars appear in the upper right hand corner of the screen.

This works well for Las Vegas. Try zooming into the Strip and using the Look Around view.

With recent images that are crystal clear, Look Around is a fun way to explore places and a useful addition to Apple Maps.

It can't fully compete with Google's Street View just yet, but it's a good start and it will be nice to see how this feature evolves.

A HANDY NEW feature in **Apple Maps** lets you share your estimated time of arrival with your friends, family or any contact.

Perhaps you are headed to your friend's house for dinner or maybe you're taking a rideshare home and want to keep your spouse posted on when you will arrive.

To use the feature, just open **Apple Maps** and begin navigation directions to a place as your normally would.

Tap the bottom of the screen where it says **arrival, min and mi** to see some more options, including a new **Share ETA** button.

Tap it to share your location, route and destination with someone in your contacts.

Your friend on the receiving end will get a text that says you're sharing your ETA with them, along with your destination and estimated arrival time. The system will automatically text them again if there is an unexpected delay along the way.

Keep in mind your contact will not be able to see your active location on a map.

ONE OF THE more eye opening tips in previous versions of this book involved all the open **Safari** tabs left behind from previous web searches and website visits.

Many were surprised to find a graveyard of old tabs simply by opening Safari and pressing and holding down the button in the lower right hand corner that looks like two overlapping boxes.

Go ahead, you can try it right now. You might see a message that says something similar to "Close All 32 Tabs." Apparently, not many people realized that unless you close them out, the sites you visit linger in the background.

With iOS 13, all of that is changing. You can now have Safari automatically close out your old tabs. It's probably a good idea to have it do so every once in a while to keep things tidy, clear out the previous websites you've visited and keep Safari from getting bogged down with lots of open windows.

To close out your old tabs automatically, go to **Settings** > **Safari** and look for the section labeled **TABS**. You'll see an option for **Close Tabs**.

Tap it to see all of your available options, which include **Manually**, **After One Day**, **After One Week** or **After One**

Month. Choose the option that works for you and back out of the menu.

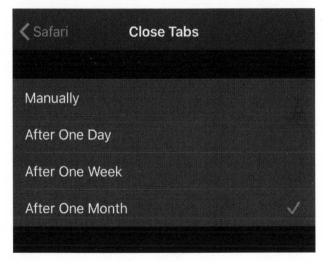

Safari can automatically close out old tabs

From now on, all of your open tabs in Safari will be closed out automatically after that time frame.

Keep in mind, closing out these Tabs is not the same as clearing your web browsing history. I'll show you how to do that in an upcoming tip.

This feature simply closes all of the websites you have open on your phone.

20 / TAP AND HOLD FOR A SNEAK PEEK

APPLE IS REALLY TRYING to make the preview a part of everyday iPhone use. This is a feature that's been percolating on the iPhone for a while now, but I'm still not convinced many people use it.

However, it can be useful when you want to get a little preview of content - like a web link - before fully committing to opening it for real.

In the past, Apple has referred to this as a "peek."

To try it, go to a website or an email message that contains a link. Instead of tapping the link to open it up, **tap and hold** your finger on it.

This will give you a little preview of the website behind the link. The website will open in its own little window on top of whatever app you're using.

You'll be able to see if it's the right link or what you're looking for. You can exit out by just tapping anywhere on the screen that's blurry.

Otherwise, you have some options to take further action with the link - you can open it up full screen, add it to your reading list, copy the link URL or share it out using another app.

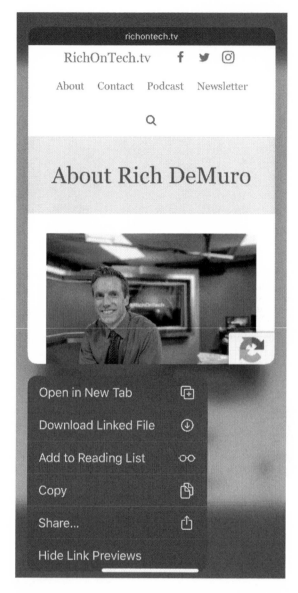

Tap and hold on a link to get a sneak peek of what's behind it

One example of how this could be useful: let's say you're sending a link to someone that someone else sent you in an email. There are several similar links and you just want to confirm that it's the right link you're emailing before you hit send.

You can use this feature to take a peek at each link, confirming the one you're sending is correct without having to open each link fully and switch back to the email app each time.

You're completing three preview actions instead of various other taps and app switches.

THERE'S BEEN a lot of talk about the iPhone battery, especially since it came out that Apple was slowing down certain older devices in an attempt to keep them running smoothly as the battery wore out.

These days, you get a lot more information and control over your battery performance. You can find this information under **Settings > Battery**.

Here, you can see which apps are using up your battery, patterns related to your phone usage and how it impacts battery life.

If you tap **Battery Health**, you'll get even more details, including the **Maximum Capacity** of your battery. The lower the number, the more your battery has worn down. This means that a charge won't last as long as it did when you first took your phone out of the box.

Under capacity, you'll see **Peak Performance Capability**, which tells you if your phone is still running at the fastest possible speed.

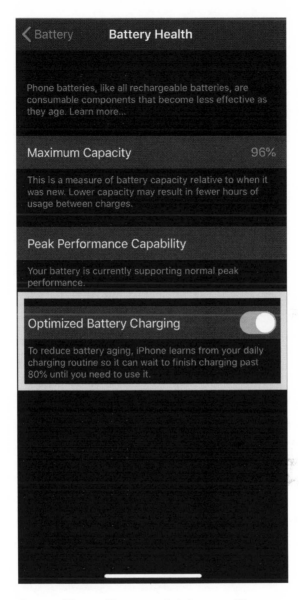

Optimized Battery Charging can help keep your iPhone
battery healthy

In many cases, it will let you know your iPhone is working at
normal peak performance, but it might also tell you that your phone
software is managing performance due to a battery issue or in a worst

case scenario, advise you to take your phone in for a battery checkup or service.

The new option to keep your battery working at its peak performance is directly below all of this. It's labeled **Optimized Battery Charging.**

By default, it should be turned on and I'd recommend leaving it this way.

Apple is applying artificial intelligence smarts here to learn how and when you charge your phone in an effort to "reduce battery aging."

What does this all mean? Apple will charge your phone up to 80% as normal, but it will wait to charge it the rest of the way until it predicts that you need your phone again soon.

This way, it's not unnecessarily wearing down your battery when you're not going to use your phone for a while.

How will this work in real life? Let's say you plug your phone in to charge before bed each night, After a bit, your iPhone will learn this pattern.

When you plug in your phone, it will begin charging and then stop at 80%. It knows that you usually wake up after 7 hours, so it will finish the charge just before you wake up. Many electric cars use similar charging smarts to optimize charge times and electricity prices overnight.

It's understandable that this setup might not work for you or your schedule. Also, if you upgrade your phone every year or two on a lease, preserving every last bit of your battery's life might not be your top priority.

In these cases, you can just turn this feature off and your phone will charge as normal.

22 / USE THE NEW CUT, COPY & PASTE GESTURES

THIS MIGHT TAKE some getting used to. In iOS 13, you can cut, copy and paste text more easily, but you have to get to know the methods first.

You know the current method, and that's not going away. And, unless you're a power user, you might never use these gestures at all. But it's my job to unlock the hidden power of your phone so here it goes.

The old method is pretty simple: select some text, then tap the copy option that appears above it on screen.

Then, tap somewhere else and look for the paste option to appear on screen. Tap it and you're done.

The new gesture based method uses three fingers. Officially, Apple describes the actions as: **pinch up to copy**, **pinch down to paste** and **pinch up two times fast to cut** the text out completely.

In my experience, you can think of the "pinch up" action as a "pinch in" action and the "pinch down" as "pinch out."

Have I lost you yet? Let's see if you can get it to work.

Open up a note on your iPhone with some text inside, preferably something that isn't very important to you.

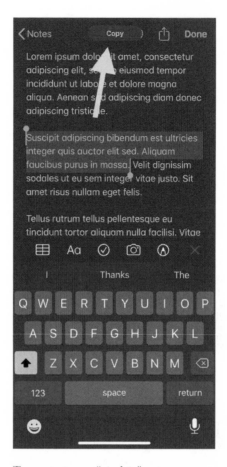

To copy text, use a "pinch in" gesture

First, select some text. Resist the urge to hit the copy button that appear above it. Instead, take three fingers, place them on the screen and then pinch them all together to the center.

Done properly, you'll see a little "copy" confirmation text appear at the top of the screen. Congrats!

Now, let's attempt to paste our text somewhere else.

Move the cursor to the bottom of your note, then place three fingers on the screen and pinch them all out. Done properly, your text will paste on the screen and you'll see a little confirmation "paste" at the top.

I'll be honest, these gestures are much easier on the iPad, where there is more screen real estate. But they work on the iPhone and once you get the hang of them, might save you a few clicks.

Ready to earn your gesture badge?

Try to cut the text completely. This time, select your text, place three fingers on the screen and pinch in twice fast. Done properly, the text should disappear.

These gestures can be tricky. Maybe there will be a "gestures" Olympic event in a few years.

I'll make getting to the next chapter much easier - just swipe or turn the page.

ARE you ready for another gesture? I promise, this one is easier.

In my last book, I explained how you can shake your iPhone to undo what you've just typed. Someone at Apple must have realized how silly that can look, because they've added **a new undo method: a three finger swipe to the left**.

Redo gets a similar treatment and you guessed it, it's just a **three finger swipe to the right**.

I can already see the song: You swipe it to the left, swipe it to the right. Turn yourself around. That's what the iPhone gestures are all about.

Ready to try it?

Start a fresh email or note on your phone and enter some text.

Now, place three fingers on the screen and just swipe them to the left. Done properly, the text you just typed should disappear and you'll see a message at the top of your screen that says Undo.

Now, place three fingers on the screen and swipe them to the right. You should see your text reappear, along with the word Redo at the top of the screen.

Told you this one was much easier. Now, remembering to actually use this gesture is going to take some practice.

Swipe three fingers on screen to the left to undo, three fingers to the right to redo

IF YOU LIKE KEEPING your apps up to date, you might be surprised at how Apple has hidden app updates in iOS 13.

Open the **App Store** and you'll no longer find an Updates tab at the bottom of the screen. It's been replaced with a new Arcade feature, which lets you subscribe to a bunch of games for a monthly fee.

So where did your app updates go?

To find them, tap the **Today** section at the bottom of the **App Store** screen. This will take you to the "front page" of the App Store.

Now, in the upper right hand corner you should see a **user profile icon**. It might even have a familiar red notification circle with a number in it. This is a signal that you have some app updates available to you.

Tap the profile icon and it will take you to an **Account** screen.

Towards the bottom, you should see a section labeled **AVAILABLE UPDATES**.

If you don't see this, place your finger in the middle of the screen and pull down to refresh the page.

Tap in the upper right hand corner of the App Store to
see app updates

To update your apps, just hit the **Update All** option at the top,
or update them one by one with the **UPDATE** button to the right of
an app name.

It's important to keep your apps up to date for a variety of reasons including better security, bug fixes and new features. Hiding the update screen like this isn't ideal, but hopefully people that don't read this book will find it out of curiosity when they see that big red dot looming over the App Store.

25 / DELETE APPS FROM THE UPDATE SCREEN

WHILE YOU'RE downloading app updates, you might see an app in the list that you no longer want or need.

Now, you can now delete it right from the update screen!

For instance, let's say you go to update your apps and you notice some random face tuner app on the list of updates. You forgot all about that app and now, you want to get rid of it.

Just **swipe right to left on an app you don't want**, and you'll reveal a **Delete** button!

One more tap and it's gone. Deleting apps you no longer need or want has never been so easy.

It's kind of sad to think of the effort developers put into making an app update, which then causes you to notice an app and think to delete it, but such is life.

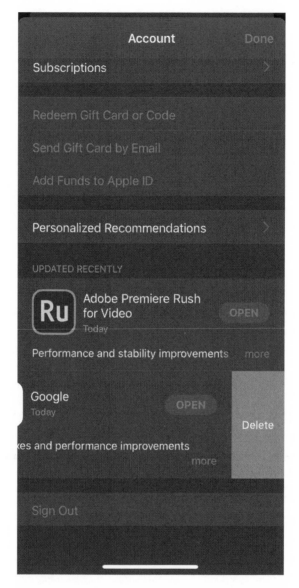

Swipe right to left on an app to delete it

I DOUBT you'll ever have an actual use for this tip, but it's a fun one for nerds like me. Plus, you'll get a kick out of using a mouse with your iPhone.

Yes, in iOS 13 you can connect a USB or Bluetooth mouse to your iPhone and control it all.

Now, before you say I'm crazy, keep in mind this is an accessibility feature, so it's there for a reason: to help everyone use an iPhone, even if they have a physical limitation that might keep them from using a traditional touch interface.

It's likely this feature will get more use on the iPad, where it makes a bit more sense to control such a large screen in this fashion.

Still, whether you need it or just want to try it out, here's how to use a mouse with your iPhone.

To start, find a mouse. Bluetooth is probably the easiest, but a USB mouse will work as well, you will just need an adapter to plug it into the port on the bottom of your iPhone.

Once you've got your mouse ready to go (in pairing mode for Bluetooth) you'll need to turn on the mouse feature.

Go into **Settings** > **Accessibility** > **Touch** > **Assistive Touch**.

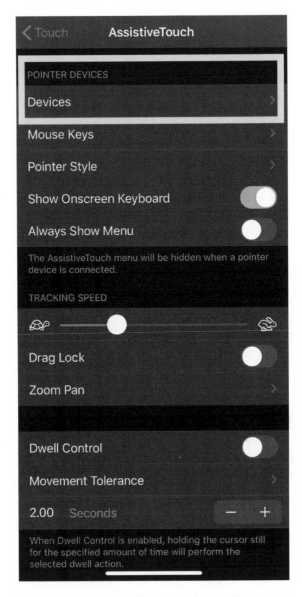

Connect a mouse to your iPhone using AssistiveTouch
options

Once you're in the **AssistiveTouch** section, toggle the feature
on by hitting the switch near the top of the page.

Now, scroll down to the section labeled **POINTER DEVICES** and tap **Devices**.

Tap **Bluetooth Devices** and if your mouse is in pairing mode, you should see it show up here. Tap to connect it with your phone.

One you do that, you should see a circular mouse pointer on your screen. Move around your mouse and it should control the pointer!

You can use the buttons on the mouse to tap around on screen. The left mouse button will select items, the right mouse button will bring up a virtual menu so you can check your Notifications, go Home, enter Control Center and more.

You can tweak this experience to your liking by using the options under **POINTER DEVICES** to change the **Pointer Style**, **Show Onscreen Keyboard**, adjust the tracking speed and more.

You can even swipe up from the bottom of the screen to reveal all of your recent apps or go home.

It is such a strange experience to use a mouse with an iPhone after all these years, but I have to admit: it's kind of fun.

But that's just the geek in me. One of these days, I need to bring my iPhone and mouse into a Starbucks and start working. I bet I'll get a lot of triple Venti takes.

PORTRAIT MODE IS a fun way to get creative with your iPhone pictures. It started with a blurry background - the pros call it bokeh - and it has expanded into much more.

There are now modes for **NATURAL LIGHT**, **STUDIO LIGHT**, **STAGE LIGHT** and more. I'm guessing you've played with some of these looks before.

In iOS 13, Apple adds a new one to the mix: **HIGH-KEY LIGHT MONO**.

If you're a Hollywood lighting expert, I probably don't have to explain what this camera mode looks like. But for the 99% of the rest us, Apple describes the effect as a "beautiful, classic look with a monochromatic subject on a white background."

I took a picture of my kid with the effect turned on and it's quite dramatic - just position their head in the on-screen circle and the background immediately melts away.

Snap a picture and Apple's algorithms go to work to process the effect further. The resulting image looks like you took a black and white picture of someone on a white background with dramatic lighting.

A photo taken in High-Key Light Mono portrait mode

To try it, open the **Camera** app and swipe over to **PORTRAIT** mode. Next, swipe left on the area where it says **NATURAL LIGHT** and scroll over to the final option labeled **HIGH-KEY LIGHT MONO**. Now, find someone to take a picture of and snap away!

I'm an instant fan of this effect since the resulting pictures look unlike anything we've seen our phones take before. Expect to see "high-key light mono" pics flood Instagram soon.

One more thing: this effect also works with the front facing camera, so try it for a cool looking selfie!

APPLE REVAMPED the **Reminders** app in iOS 13 to make it much more powerful, especially at understanding what you're trying to say.

You can now type in a reminder in natural language and the app will figure out the dates, times and more.

For instance, if you write:

Call the school office Friday at 9:30 AM

You'll see a suggestion above the keyboard to change Friday at 9:30 AM to Friday's date. Another tap and the Reminders app will convert the time into a notification for the alert to go off at 9:30 AM.

Alternatively, you can start with a time or location for your reminder. Just tap the **clock icon** above the keyboard to see a few dates in the near future to tag your reminder with. The **location arrow icon** will bring up an option to fire off your reminder when you are **Getting in Car**. Or, hit **Custom** to have your reminder go off at a place you choose or when you get out of the car.

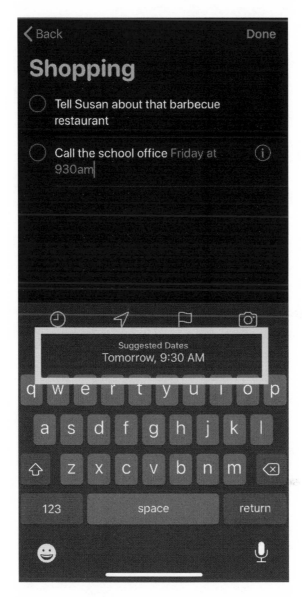

Use natural language to write reminders and the app
will figure out the rest

For instance, you can set a reminder to go off the next time you're
in a specific store. Let's say you need wood glue from Home Depot.

Just search for Home Depot and choose a location. The next time you enter that store, Reminders will send you a notification.

If you've already created a reminder and want to attach a location to it, just select the reminder and hit the button with the **i** in a circle.

Now, you can customize the option to **Remind me at a location**.

Elephants will have nothing on you.

29 / ADD AN ATTACHMENT TO A REMINDER

APPLE'S **REMINDERS** app has been pretty sparse over the years, but it finally gets a feature that really upgrades its usefulness: the ability to attach items including scans, photos and web links to a reminder.

Sometimes, a picture is worth a thousand words. This feature makes it easy to attach an invitation to an RSVP reminder or a photo of a part number you need to pick up at the store.

To add a photo or document scan, open the **Reminders** app and start typing a reminder. When you're ready to insert, tap the **camera icon** above the keyboard.

You'll see options to **Take Photo**, **Photo Library** or **Scan Document**. Choose the one that works for you, then hit **Done** when you're finished. Your image or document will show up in the list of reminders.

If you want to add a web link to a reminder, you can now do that too. Before, you could add a link but it was not "clickable," which diluted some of the benefits of the link. You would have to copy and paste it into a web browser to actually access it.

Adding a link isn't as straightforward as you might think. If you simply copy and paste a web link into a reminder, it still won't be clickable.

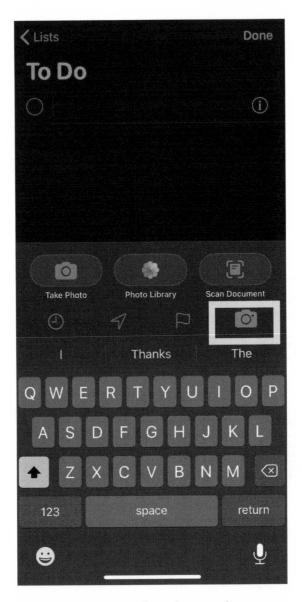

Tap the new camera icon in Reminders to attach
pictures and scans

The way to do it is to create your reminder first, then hit the
circle **i** button to the right of it. In here, you'll see an option for **URL**.
Paste your link here and hit **Done** when you're finished.

The reminder will now show a little preview of the link, which is, indeed, clickable.

ANOTHER NICE TOUCH in the Reminders app is the ability to set an icon to quickly identify your list.

Previously, you could only customize the color of the list title. Now, there are more colors to choose from and you can choose an icon to go along with your list.

There are icons for bookmarks, birthday presents, wallet, working out, medicine, wine and more.

To create a new list with a custom look, open the **Reminders** app to the main screen and hit the option in the lower right hand corner to **Add List**.

From here, give your list a title and hit **Done** on the keyboard when you're finished.

Now, you can see all of your options for color and icons.

Tap a color and the icon instantly changes to match.

Tap an icon and the list icon instantly changes.

Once you have the right combination of color and icon, hit **Done** in the upper right hand corner and your list is ready to go on the main screen.

Customize the look and feel of your lists in Reminders

To customize a list you've already created, just tap it from the main screen under "My Lists" and then hit the button with **three dots** in the upper right hand corner.

This will bring up a few options including **Name & Appearance**. Tap it to customize the name, color and icon of your list and hit **Done** when you're finished.

EVER SAY to yourself "remind me to tell Susan this the next time I talk to her." Well, if you're not friends with someone named Susan, probably not. But you get the idea.

Reminders has a nifty new feature where you can set a reminder to notify you of its contents the next time you start text messaging someone specific.

For instance, let's say you want to tell someone that you tried the restaurant they suggested and it was delicious. Maybe it's too late at night to text them right then, or not pressing enough to warrant a conversation right this very second.

First, open the **Reminders** app and create a reminder to **Tell Susan about that BBQ restaurant**. Then, hit the button to the right of it - the one that looks like an **i** in a circle.

In here, you'll see an option for **Remind me when messaging**. Toggle it on and then hit the option to **Choose Person**. Choose the associated person from your contacts list and hit **Done** when you're finished.

The next time you start messaging that person, your reminder will pop up on your screen. You'll never forget to tell someone something ever again.

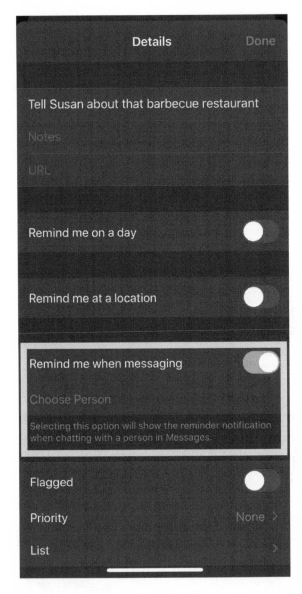

Attach a person to a reminder and it will pop up the
next time you message them

Why not set a reminder right now to tell your best friend how
incredibly useful this book is the next time you message them?

HERE's an example of a desktop computer feature making its way to the iPhone. In the past, if you wanted to get to the end of a long article, document or webpage on the iPhone, you would have to continue flicking your finger on your screen until you made your way to the bottom.

In iOS 13, you can now drag the scroll bar that appears on screen to quickly get to the bottom (or back to the top) of a document.

To try it, go to a website that has a lot of text on it, like a long newspaper article. If you can't think of a site to go to, try longreads.com. They have a bunch of long articles perfect for trying out this little experiment.

Once you have your long page ready to go, start scrolling the way you normally do, by tapping your finger on the screen and scrolling or flicking to move the document.

Now, notice on the right hand side of the page there is a faint **vertical bar** that appears as you scroll down - this is the scroll bar.

Tap and hold the scroll bar until you feel a faint vibration - this means you've grabbed it and it's ready to scroll.

richontech.tv

Prior to joining KTLA, Rich was a Senior Editor at the technology website CNET and worked as a reporter at Channel One News and local TV stations in Yakima, Washington and Shreveport, Louisiana.

Rich is originally from New Jersey and graduated from the University of Southern California with a degree in Broadcast Journalism. He lives i Los Angeles with his wife and sons a u enjoys traveling, reading, running, magic, movies, music and writing.

Hard press, then drag the scroll bar to move up and down a page fast

Without lifting your finger off the display, move it up and down the screen. You'll move the document up and down faster than ever.

I know, I know, it's the little things in life.

Keep in mind, the scroll bar only appears when you start to scroll through a document, and it goes away fast. Too fast, in my opinion. This means you only have a little window of time to grab it.

Now, rock 'n scroll!

YOUR PERSONAL HOTSPOT is about to get much more useful in iOS 13.

Here's the old routine: turn on the hotspot feature on your phone, then go to the WiFi settings on your other device, like an iPad, wait for it to see the hotspot network, tap it and wait for them to connect.

In iOS 13, all you have to do is toggle one setting and your other device will automatically connect to your personal hotspot if there's no other internet connection available.

Example: let's say you have an iPhone and an iPad. You travel a lot, and most of the time you're connected to the hotel Wifi, but there are times when you don't have a connection.

Without you doing a thing, as long as your iPhone and iPad are near each other, they will link up so you can use the internet connection from your phone on your tablet.

To turn on this feature, head into **Settings** > **Wi-Fi** on your other device, like an iPad.

Then, look for the option labeled "**Auto-Join Hotspot**." Tap to see your options, which include **Never**, **Ask to Join** and **Automatic**.

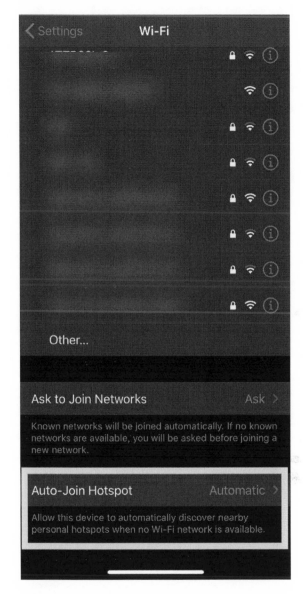

You can have your iPhone automatically join personal hotspots

Never means that your tablet will just ignore the fact that your iPhone is nearby and it can tap into its internet connection.

Ask to Join will remind you that you can connect to your iPhone for internet since there is no other network to connect to.

Automatic will just seamlessly acknowledge that your iPad has no internet and your iPhone is nearby with a shareable connection. It will use your iPhone's hotspot to connect to the internet.

Choose the best option for your data plan - if it's more limited, you might want to use **Never** or **Ask to Join**. If you have a good amount of hotspot data allotted to you each month, **Automatic** sharing will make your life much easier if you roll with several Apple devices in your bag.

SHARING your iPhone's personal hotspot with family members is also easier in iOS 13.

Instead of manually having to link up and connect your family's devices to your hotspot every time they want to use it, you can now just enable the connection automatically.

Apple calls the feature **Personal Hotspot Family Sharing**, and if you've ever used any other aspects of family sharing, it works in a similar way.

To start, your family members must be identified as such on your iCloud account. If you haven't set this feature up, go to **Settings** > **Apple ID** > **Family Sharing**.

Once that's done, go into **Settings** > **Personal Hotspot** and tap the option for **Family Sharing**.

Toggle the switch on and you should see your family members appear below.

You can now choose the appropriate permissions for each family member: whether they have to **Ask for Approval** each time they want to use your hotspot, or make the connection **Automatic**.

The choice is up to you, but if you want to make your life easier, **Automatic** is the way to go.

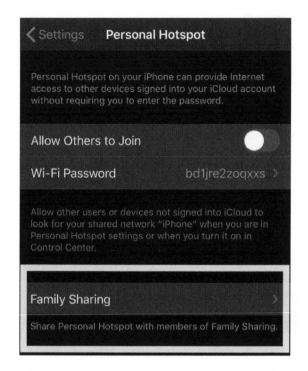

Family devices can take advantage of your phone's
hotspot automatically

Here's an example: let's say you're out to dinner with the kids and
they are allowed to use their iPads during dessert.

With the old way, you'd have to turn on your iPhone's hotspot,
then go into the WiFi settings on their iPad, wait for your Personal
Hotspot to appear, and select it from the list to connect.

With Family Sharing enabled, the connection would be made
Automatically, or if Ask for Approval is selected, you would get a
prompt on your iPhone that the iPad wants to make a connection.

Either way, it's a faster method to establish the connection.

IF YOU'RE on a super limited cellular plan or just want to keep most of your data usage on WiFi, iOS 13 has a new feature that can help.

It's called **Low Data Mode**, and once you turn it on, it helps the apps on your phone reduce their cellular data usage.

What does this mean? Apps won't refresh their information in the background or do things that aren't necessary at that exact moment.

To turn it on, just go into **Settings** > **Cellular** > **Cellular Data Options**. You'll see a toggle for **Low Data Mode**. Turn it on if you want to restrict the amount of data apps use, especially in the background.

Keep in mind, developers must code their apps so they play nice with this feature. Not all apps will be on board with Low Data Mode. If they're not made with this mode in mind, they might continue to use data freely.

Still, if you're concerned about high data usage and you're on a strict cellular data budget, this is a welcome option.

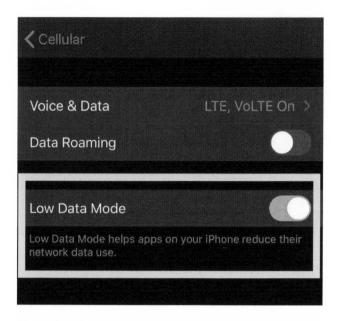

Turn on Low Data Mode so apps use less of your cellular data

Think of it this way - once Low Data Mode turned on, apps will only do the things you want them to do when you're on a cellular connection. They won't be secretly using data in the background, which can save you many precious megabytes.

Do Not Disturb is a great feature on the iPhone. It enables you to sleep or take a break from your device without being disturbed by incoming calls or notifications. At the same time, it allows the important contacts you designate to still ring your phone.

You can set up **Do Not Disturb** to follow a schedule, but you can also enable it temporarily.

To do this, enter **Control Center** by swiping down from the upper right hand corner of your phone (swipe up from the bottom of the screen if you have a home button) and look for the **crescent shaped moon icon**. Tap it to enter Do Not Disturb mode and tap it again to exit the mode.

Easy on and easy off.

But, you can also tap and hold the the Moon icon for additional Do Not Disturb options. These include **Do Not Disturb For 1 hour**, **Until this evening**, **Until tomorrow morning** and **Until I leave this location**. Your options change depending on where you are and what time it is.

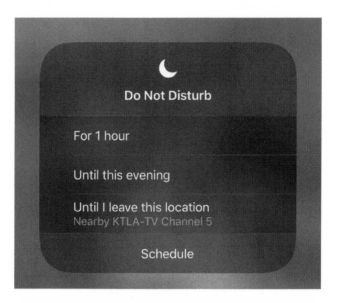

You can get more specific with Do Not Disturb with a long press

New in iOS 13, **Until I leave this location** shows your current location so you know exactly when the feature will flip off.

To select one of these finer Do Not Disturb controls, just tap it. You'll see that the moon icon is now lit up, signifying that Do Not Disturb is now in effect.

If you want to tweak your Do Not Disturb settings, just long press the Moon icon again and hit the **Schedule** button at the bottom.

You can now change everything from who to allow calls from to whether you want to allow repeated calls to ring your phone in what might be an emergency situation.

Many people are hesitant to set up Do Not Disturb because they are worried they might miss an important call, but if you take the time to set it up properly, you'll find it can add a little zen to your phone routine.

In iOS 13, you can finally share audio to two different devices - specifically AirPods, at least for now.

If you and a friend want to watch a movie on the train together and both have a set of AirPods, you can send audio simultaneously to both AirPods at once.

To set it up, you'll need two pairs of AirPods. I'll wait here while you go buy some.

Just kidding.

First, ensure both AirPods are paired with the iPhone. You can confirm this by going into **Settings** > **Bluetooth** and confirming that both sets of AirPods are listed under **MY DEVICES**.

If your friend's AirPods are not, have them open their case and pair them with your phone as usual.

Once you see both pairs connected, play some music or a video or whatever you want to share listening to.

Now, enter **Control Center**. You can do this by swiping down from the upper right hand corner of the screen (or if you have an iPhone with a home button, swipe up from the bottom of the screen.)

Once you're in Control Center, look for the box in the upper right hand corner that contains your audio controls. You should see a snippet of the name of the audio you're playing along with a play button and another icon that looks like a circle with a triangle pointed to the center of it.

Tap that icon.

To start sharing audio, tap this icon

This will bring up your audio routing controls. You should see a section named **HEADPHONES** with your AirPods listed under there.

To select multiple AirPods to listen on, tap the **circle** next to the name of each pair. The circle should display a checkmark in the center, signifying that those devices are now receiving audio simultaneously.

To disconnect a device, just hit the **circle checkmark** once again.

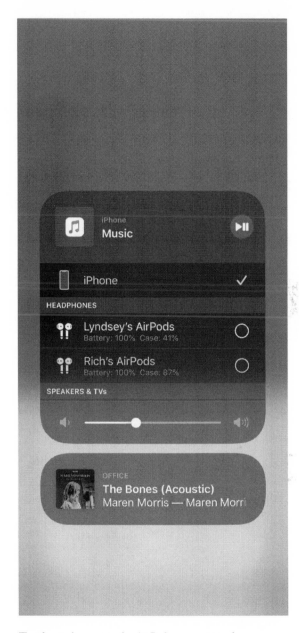

Tap the circles next to the AirPods you want to share
audio between

This feature got one of biggest responses at WWDC 2019 when

Apple announced it. It's kind of like having a super expensive head-phone splitter built into your iPhone.

Once both sets of AirPods are connected, you'll see a new "people" icon on the Volume switch in Control Center.

Long press the Volume switch to bring up two separate volume controls for each set of AirPods.

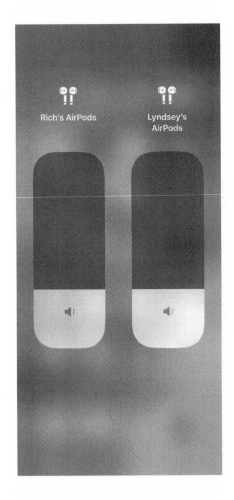

Right now, the feature is limited to Apple branded headphones, which includes AirPods, AirPods Pro and some Beats models.

Perhaps Apple will expand it to include other compatible Bluetooth headphones in the future.

If in doubt, just pair both headphones to your iPhone and see if the circle option is there.

At the time of this writing, I tested a pair of non-Apple headphones and the audio sharing feature did not work.

APPLE GAVE the share sheet a major overhaul in iOS 13. While the basic functionality remains the same - to share your content to other people, apps and devices - the look is brand new.

In case you're wondering what the share sheet is, it's the menu that pops up when you hit the **Share** button. That's the one that looks like a box with an arrow pointing out of it.

It's the menu you use when you want to share a photo, web link or just about anything else on your phone.

Before, the entire share sheet was icon oriented. Now, it's a combination of icons and menu items in a list.

Although the share sheet seems to be using a bit more artificial intelligence to figure out who and how you want to share a particular item, you can still customize a lot of what's displayed here.

To see what I mean, open up a photo and hit the **share** button under it.

You'll see some suggested people to share with, then a row of app icons, followed by more options in a menu format.

Scroll the app icons over to the left until you see a **More** icon and tap it.

From the Share Sheet, scroll over and hit More to customize which apps show up

Now, hit the **Edit** button in the upper right hand corner.
You can now customize what shows up here by pressing the **plus**

or minus icons next to items. You can also turn off some items completely by toggling the switch next to them.

Finally, once you are done adding and subtracting items from your Favorites, you can use the **three little lines** to "drag" the icons in the order you like.

Hit **Done** twice and you can see your handiwork. Keep in mind, you will still see Apple's suggestions after your favorites if you didn't turn them off completely. For some apps, the option to turn them off completely isn't available.

Next, you can work on the menu items below using the same process.

Scroll all the way down to the bottom of the share sheet and hit **Edit Actions...**

This will take you to a similar screen where you can pick and choose your **Favorites** and organize them in the order you like.

Hit **Done** when you're finished.

Now, a note about how Apple has changed the share sheet. It's much more dynamic, offering up options that are relevant to what you're trying to do. So the Photos share sheet will be different from the one you see in Safari.

But the biggest difference this time around is that you have your Favorites and Apple's suggestions. Your favorites always appear first in the list, followed by the rest of what Apple thinks might also be helpful.

In iOS 13, Find My Friends and Find My iPhone app have merged, creating one new mega-app called **Find My**.

This app can help you find your friends and devices as the name implies, but there's also a new option to help you locate your devices even if they're not connected to the internet.

To set it up, open the **Find My** app and you'll see three options along the bottom.

People contains your friends and family members with Apple devices that you can share your location with. Tap **Share My Location** and type in a phone number or email address associated with an Apple account to share your location with this person for one hour, until the end of the day or indefinitely.

If someone is sharing their location with you, you'll be able to see them on the map. But there are many more useful features under the hood if you tap someone's name. Try tapping **Add**... under **Notifications**.

This will bring up a new option to **Notify Friend**. You can set up location based notifications which are useful for tracking family members.

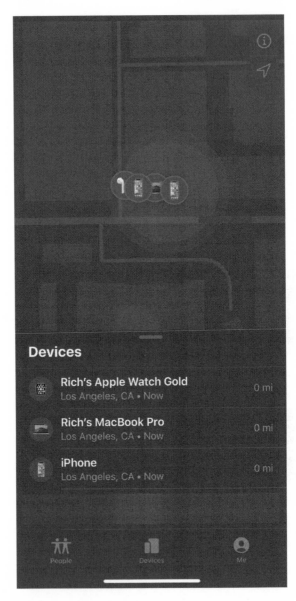

See where people and devices are in the new Find My
app

For instance, you can set this up on your child's device so you will
get a notification every time they leave or arrive at school, or any location you choose.

Back on the main screen, the next tab is **Devices**. This gives you a quick glance at the location of all of your connected devices.

Tap a device to get more info including the battery level, along with options to **Play Sound** (helpful for locating your device under a couch cushion in your house) and **Directions** to your device.

Scroll a bit and you'll find the useful option to **Mark As Lost**. This will display contact information on the lock screen of your device so if someone finds it, they can get it back to you.

I did a TV report on a guy who lost his Apple Watch bodysurfing in the ocean and it washed up on the shore a few months later. The person who found it charged it up, saw the owner's phone number on the screen and got in touch with him to return it. Also, it still worked.

The other option is to **Erase This Device**. This is useful if you lose your device and don't think you're going to get it back. The device will attempt to erase itself the first time it is connected to the internet. It's a last ditch effort to protect your personal information if your device ends up in the wrong hands.

One more important website to know: icloud.com/find.

This is the place to go if you lose your device, or if a friend loses theirs. You can log into this website with your Apple ID and access the same tools to locate your friends or devices.

THERE HAVE ALWAYS BEEN a lot of apps available for the iPhone that can help you edit videos, but now there are helpful tools built right into the **Photos** app.

Where before you could just trim the ends off of a video clip, now you have many more functions available to you including the ability to adjust the color, contrast and other visual aspects, apply filters and even rotate and crop video.

Basically, if you've ever edited a photo on your iPhone, you can apply nearly all of the same edits to a video.

The best part: nearly all of your edits are non-destructive to the original video. This means if you ever change your mind, you can get the original back in just a few taps.

To try it, open up the **Photos** app and select a video, then hit the **Edit** button in the upper right hand corner.

This will bring up all of your editing tools.

For starters, there's a **Volume** icon in the upper left hand corner. Tap it to mute the volume on your video. This means when you share it, it will no longer have any sound attached.

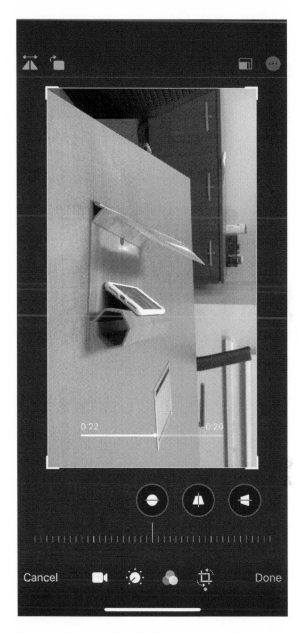

Rotate a video, mute sound and more with new video
editing tools

On the bottom row, there are buttons for **VIDEO**, **ADJUST**,
FILTERS and crop and rotate tools.

The **VIDEO** button lets you mute the audio and drag the sliders on the video thumbnail to adjust the in and out points.

ADJUST lets you automatically adjust the visual look of a video with the magic wand button, just like you would on a picture. Apple's software will automatically apply edits that it believes will make your video look its best when it comes to exposure, highlights, shadows, contrast, brightness, black point, saturation, vibrance, warmth, tint and more. There's also sharpness, definition, noise reduction and a vignette tool.

At this point, you're probably thinking that you will never use all of these tools, and you're probably right, unless you're a video pro or social media superstar. Still, it's great to have these tools built into your phone if you ever need them.

To manually adjust a setting, tap one of the buttons, then swipe the slider below left or right. You'll see a real time representation of the changes. Tap the video to compare your changes to the original.

When you're happy with your changes, you can hit **Done** or **Cancel** to back out.

Next to **ADJUST** is **FILTERS**, which apply looks that you might have used on your pictures over the years. Think of these as Instagram filters, but for your videos.

The final option is a button that brings up crop and rotate tools.

These can be really handy for resizing your video or flipping it horizontally or vertically. It will also rotate your video so it's oriented the proper way if you mistakenly recorded something sideways.

Starting at the upper left hand corner is an option for vertically flipping your video. Next to that is a rotate button and to the right is a crop tool.

You can crop to any size you want, or apply common crops including **SQUARE**, **9:16**, **3:4** and more. Try them all out and if you do something you don't like, just hit **RESET** at the top or **Cancel** at the bottom.

De-select the crop tool and you have three more buttons at the bottom including **STRAIGHTEN**, **VERTICAL** and **HORI-ZONTAL**. You can use these to adjust the look of your video even more.

When you're finished fixing your video, just hit **Done** to save your work. Keep in mind you can come back into your video at any time and hit the **Revert** button that now shows up when you go into **Edit** and get back the original video.

These tools are a welcome addition to the Photos app and can be super helpful not only for the occasional video that you record in the wrong orientation, but they allow you to be more creative with your social media videos without downloading any additional apps.

As a journalist who has been working with video for many years, it's amazing to see tools this powerful on a smartphone. These are edits that used to take highly specialized equipment and know-how.

Now, they're built into pockets everywhere.

41 / PROHIBIT WEBSITES FROM ACCESSING YOUR CAMERA & MICROPHONE

APPLE HAS ADDED finer controls for Privacy when it comes to websites trying to access your Camera and Microphone.

Previously, Safari grouped together Camera & Microphone Access into one option that you could simply toggle on or off.

Now, these options are singled out and there's a new one added to the mix: Location.

You can have each website ask permission for access, deny access to all websites or allow blanket access, which I definitely don't recommend.

The default is **Ask**, but you are in charge here. If you want, you can deny all websites access to your Camera, Microphone and/or Location.

It might be a good idea to deny access to Camera and Microphone, but Location is useful for websites to have access to for a variety of reasons.

To change these settings, go into **Settings** > **Safari** and look for the section labeled **SETTINGS FOR WEBSITES**.

You'll see options for **Camera**, **Microphone** and **Location**. Tap to see your choices for each, which are **Ask**, **Deny** and **Allow**.

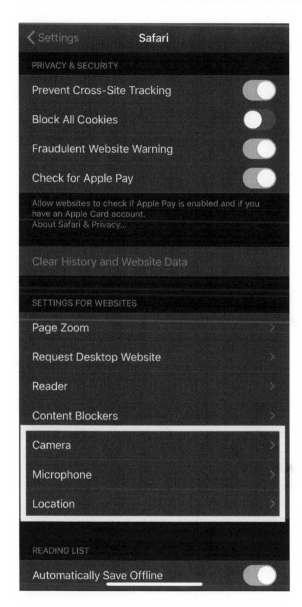

Prevent websites from accessing your phone's camera,
microphone and location in Safari settings

Tweak these to your liking and if you ever change your mind, or come across a website that doesn't seem to be working properly, remember that you changed these settings.

IF you previously set up **Screen Time**, you might have noticed that while you could set up a time limit for a group of apps, say Social Networking, setting up a time limit for a specific app wasn't possible.

Until now. iOS 13 now lets you set a time limit for specific apps. If you want to limit yourself to just 30 minutes a day for Facebook and 1 hour a day for Instagram, you can now do just that.

To set it up, go to **Settings** > **Screen Time** and turn on the function if you haven't already done so.

Now, tap the section that says **App Limits** and hit **Add Limit**. Here, you will see all of your app categories.

To choose a specific app to limit your time on, just find it in the appropriate category. You might have to browse around a bit to locate it.

Even easier, you can reveal the hidden Search bar by "pulling down" on the page itself. Just pull down from the center of the page ever so slightly to reveal it. If you pull too hard, you'll exit this screen completely.

Now, search for the name of the app that you want to set a limit for. Let's say it's Twitter. Search Twitter, then hit the circle next to the name of the app to select it.

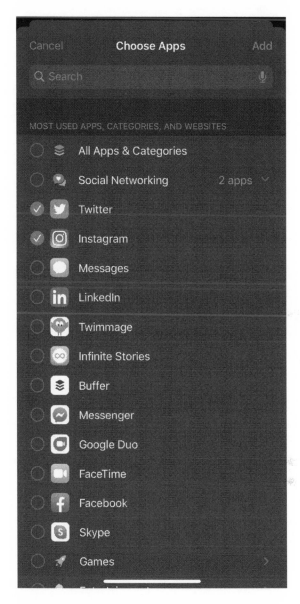

Set time limits for specific apps in Screen Time

Hit **Add** and you're ready to set a time limit.

You'll notice that Twitter is now selected on the **Choose Apps** screen.

Hit **Next** in the upper right hand corner and you can now set a time limit.

Want different time limits for each day? You can tap **Customize Days** below the time wheel to choose them. Finalize your selections by tapping **Add** in the upper right hand corner of the screen.

When you're about to hit your time limit on an app, you'll get a five minute warning that time is almost up. When it is, a giant Time Limit message will take over your screen, along with the option to run under a rock and hide (OK) or Ignore Limit. Tap Ignore Limit for new options including One more minute, Remind Me in 15 Minutes or Ignore Limit For Today.

You are an adult, so you are in control, after all. But if you set up these limits for your kids you can make it so they have to send you a request for an extension.

43 / SET A TIME LIMIT FOR A WEBSITE

NEW IN iOS 13 is the ability to set time limits for specific websites. Let's say you spend way too much time researching vacations on TripAdvisor and you want to limit yourself to a comfortable 20 minutes each day.

Go to **Settings** > **Screen Time** and tap **App Limits**.

I know, not the best description for a place to go to limit websites, but that's how they labeled it.

From here, tap **Add Limit** and look all the way at the bottom of the list for the option labeled **Websites**. You might see a list of suggested websites. You can choose one of these or tap the **Add Website** option at the bottom to type in your own.

For our example, we'll type in tripadvisor.com. Hit **Done** when finished. Continue this step if you want to add more websites to the list until you're finished. When you are, hit **Next** in the upper right hand corner.

Choose a time limit, then hit **Add** to finalize your selections.

Now, you can browse to your heart's content on these sites, but only up until the limit is reached.

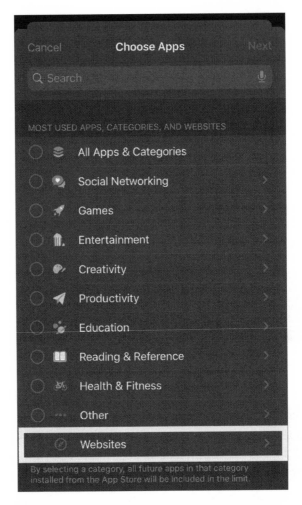

Set time limits for specific websites by entering them here

With websites, there is no extending your time. If you still want to browse, you'll have to go in to Screen Time settings and up the time limit or delete it completely.

44 / SHARE A PHOTO WITHOUT REVEALING ITS GPS LOCATION

PRIVACY IS top of mind these days. There is a welcome new feature in iOS 13 that helps protect yours when you share a photo.

Now, when you go to share a picture, you have an option to strip away the secret location data contained inside that photo.

You're probably aware that your camera attaches the GPS location to the pictures you take, which is handy for organizing your pictures and searching for the photos you took, in say, Hawaii.

But when you share a photo, that GPS data normally goes with it, which means the person or service on the receiving end can see exactly where you took it.

No big deal if you're sharing a cute picture of your kid with grandma, but if you're uploading a picture to a dating website or texting a picture to someone you're hoping to sell an item to, they might be able to see exactly where that photo was taken.

This could reveal your home address, where you work or another location you might want to keep private.

A good example of this is when you upload a photo to Instagram days after you've taken it and somehow it still knows the original location where you took the photo. Instagram is looking at the GPS data contained in the photo's file to figure out where it was taken.

Tap Options when sharing a photo to strip GPS
Location data

To see how this works, select a photo from your Camera Roll and
hit the **share** button. At the top of the screen, you'll see some text

that says **Photo Selected**, along with the location where it was taken. Tap **Options**.

In here, you'll see a section labeled **Include**. You have several options including **Location** and **All Photos Data**.

Toggle the **Location** switch off and GPS coordinates will be stripped from the photo before it's sent, which better protects your privacy.

Once you do this, when you return to the previous screen, your photo should say **No Location** at the top.

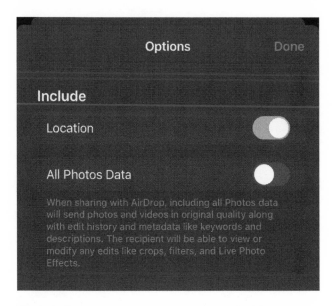

Toggle the Location switch so recipients can't see GPS data in a photo you share

Back in the Options screen, the **All Photos Data** is an interesting one. This pertains to photos you AirDrop and if you leave the toggle on, your recipient might be able to undo any edits you've made to a photo.

Keep in mind these are one time selections. You'll have to make new selections every time you share a photo.

A good rule of thumb is that it's probably OK to leave some of this data in there if you're sharing with a family member or close friend, but don't include it if you're sharing with a stranger or someone you don't know well.

45 / TURN OFF AUTO-PLAY FOR
VIDEOS & LIVE PHOTOS

PHOTOS GOT a total revamp for iOS 13, and one of the new features brings your photos to life as you scroll past them.

If you've had Live photos turned on in the past, you've probably noticed as you scroll through your camera roll the photos sort of coming to life and then stopping on the frame of your picture.

In iOS 13, your photos and videos come to life more often, and in more places. You'll notice videos playing and live photos animating as you scroll through the various ways to look at your pictures in the Photos app.

For some, this might be a welcome way to re-live old (or recent) memories, but for others all the busyness might be distracting or annoying.

To turn off the Auto-Play feature, go into **Settings** > **Photos** and look for the section labeled **PHOTOS TAB**.

Under it, you'll see an option for **Auto-Play Videos and Live Photos**.

Toggle it off and your pictures and videos will stay still unless you specifically play one of them or tap and hold on a photo to see the live version.

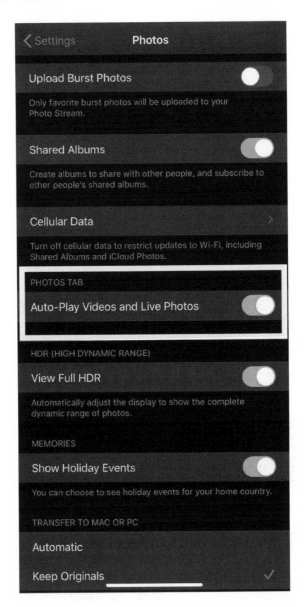

Too much movement in the new Photos app? Toggle
Auto-Play off

Toggle it back on if you want more action as you browse through
your pictures and videos.

SEARCHING FOR PEOPLE, places and things inside your photos is easier than ever in iOS 13.

Open the **Photos** app, tap the section labeled **Search**.

You can now type in a keyword and see pictures related to that term. It might be a place where you took the picture, someone in that picture or even an object in the photo.

Your iPhone will populate the Search screen with suggested items to search for, but you can also play around in the search box to see what might be in your photos.

I like to start by typing in the alphabet letters one by one. For instance, if I put an "a" in the box, I get Los Angeles, Art, August and Autumn.

"B" brings up Beach, Brooklyn, Basketball and Breakfast.

Fun, right?

Photos search is even better in iOS 13, so if you haven't used it in a while, you might want to revisit it.

And the next time someone asks you to show them pictures of your recent trip to Hawaii or the amazing ice cream cone you're talking about, head to search first to find the photos, fast.

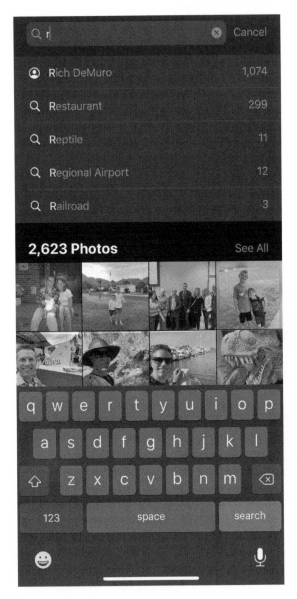

Use Search to find specific photos quickly

Since privacy is always top of mind these days, you'll be happy to know that all of this image processing is being done on your phone. Apple isn't sending your photos to some server to identify what's in them. The image recognition is being done on your iPhone and the index is stored on your phone, too.

You will notice that once you search for something, there is a new section created labeled **Recently Searched**. You can hit the **Clear** button if you want to clear your recent searches.

WHILE YOU COULD ALWAYS ADD a helpful note to a calendar item in iOS, you couldn't actually attach a document to an event.

That's all changed thanks to a new option in Calendar called **Add attachment...**

This is going to be a lifesaver for folks who like to attach a picture, document, map clipping or anything else related to a calendar event that isn't copy and pastable.

I'm a big fan of attaching the original invite of events I go to so I have all the information I need on hand. Sometimes invites are a JPEG or PDF, which means I would have to painstakingly retype all of the information contained inside into a calendar event.

Now, when you're creating or updating a calendar item, just look for the option for **Add attachment** and you can attach all sorts of files.

Attaching things like PDFs, spreadsheets and other items you've downloaded or saved to your iPhone or cloud services is easy. Just browse to find what you're looking for and tap to attach.

For some reason, Apple has decided that photos aren't as important to calendar events, so attaching one will take an extra step.

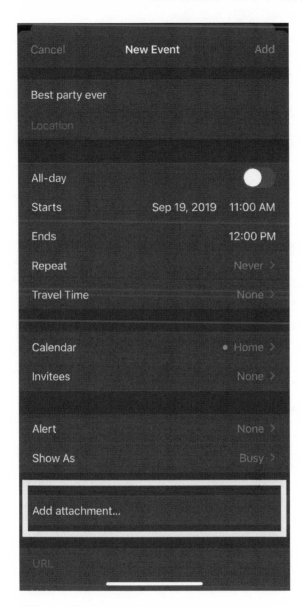

Add an attachment to a calendar event

First, go to the **Photos** app and find the picture you want to attach. Tap the share icon under the photo (the one that looks like a box with an arrow coming out of it) and scroll up until you see the option for **Save to Files**.

Tap here and decide where you want to store the photo in your file manager, then hit **Save** in the upper right hand corner when you're finished. Be sure to remember the location of the folder where you stored it.

Now, go back to your calendar event, hit **Add attachment...** and locate your file. Tap it once to attach it and you'll now see it included in your event.

If you don't see the option for Add attachment, it might be that the calendar you're using doesn't support it. I tested the feature with an iCloud calendar and it works, but my Google calendar did not.

When you're finished adding all of the details of your event, hit **Done** in the upper right hand corner. You'll now see your attachment along with all of the typical calendar event details. Tap the attachment for quick access to the file.

Keep in mind, you can add several attachments to a calendar event.

There is probably some limit to this number, but I've added four and haven't hit it yet. Go crazy, you calendar rebel you, and let me know if you reach it.

MESSAGES HAS HAD search for a while, but it's generally hidden. You can reveal the search bar by going into **Messages** and **pulling down on your list of texts**.

There it is! Not sure why Apple hides these things, but they do.

Now that you know how to find it, let's take a look at some of the new ways you can use it to find just what you're looking for hidden amongst all of your messages.

For starters, tap into the **Messages Search** bar to bring up a new Search screen.

At the top are some people you frequently or recently messaged with. Tap one of them to go right into your conversation with them.

Under these people, you'll see a heading for **Photos**. Here, you'll see some recent photos exchanged in your messages. You can tap one to be taken straight to that message chain. Tap and hold one of these pictures and you'll get options to Save it, Copy it or Share it.

There's even more if you tap the **See All** link in the Photos section. This will take you to a gallery that contains all of the photos and videos you've exchanged through texts. You can even narrow down to just Screenshots by tapping the toggle near the top of the screen.

Search in Messages lets you see the things you've
exchanged including Links and Photos

Again, you can tap an image or video once to be taken its conversation, or tap and hold an image to get your Save, Copy and Share
options.

Another thing you can do inside Search in Messages? Search using a keyword or two.

Keep in mind, this is not Google, so it's not perfect. Start with a keyword you remember in the conversation and narrow it down from there with maybe a second keyword.

As you type in a keyword, Search will bring up the appropriate conversation and highlight the word in a snippet of the text message. Tap a message in the list to be taken to the part of the conversation where the keyword(s) appear.

You'll notice that the text on your screen that contains the keyword flashes for a bit, but blink and you might miss it.

49 / SELECT A TIME WHEN ALL DAY REMINDERS POP UP

BY DEFAULT, your iPhone will send you a notification containing your "all day reminders" at 9 AM.

For instance, if you just said to Siri on your phone, "remind me to pick up milk tomorrow," she would set a reminder with the default time of 9 AM and send you a notification at that time to remind you to do that.

You could go in and change the time on that individual task to something more suitable, but now there's a better option.

In iOS 13, you can now customize the default time that these "all day notifications" will show up on your screen.

There are many reasons for changing this behavior, but many times, you might want to have them chime when you first get into the office, or when you get home and start completing your errands.

Whatever the reason, to change the default time, go into **Settings** > **Reminders** and look for the section labeled **ALL-DAY REMINDERS**.

Toggle it on if it isn't already, and you can now select a custom time to display notifications for all day reminders.

Again, this will apply to reminders that you didn't necessarily assign a specific time to when you created them.

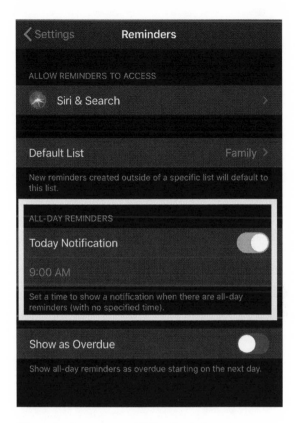

Choose the time you want All Day Reminders to pop up on your phone

ON YOUR DESKTOP COMPUTER, you probably have a folder where downloads from the web are saved. These could be pictures, PDFs and the other assorted items you download from the internet as you go about your day.

On the iPhone, you can now set a similar folder of your choice. This means that you now have one set place on your phone that contains all of the items you download in Safari. New in iOS 13, it can be a location of your choosing.

This means it will be much easier to find downloads down the road, since you can look inside the folder you set.

Come to think of it, "Downloads Down the Road" should be the name of my band, if I ever start one.

To set a download folder, go into **Settings** > **Safari** and look for the section labeled **GENERAL**.

In here, tap the option for **Downloads**.

Now, choose where you want your downloads stored. Basic options include **iCloud Drive** or **On My iPhone**.

Decide where you want Safari to store your downloads

If you use iCloud, especially in coordination with a desktop computer, this could be a good place since anything you download to this folder will immediately be synced to the same folder on your computer.

But the options are limitless. You can tap **Other**... to select a wide range of folders and destinations. Make it your own, and never lose track of a download ever again.

IF YOU'VE AMASSED a large quantity of Notes over the years, you'll love a new feature that lets you see everything attached to your Notes all at once.

This can be handy for finding something you know you attached to a note at one time, but can't necessarily figure out which one it's in. It's also a visual way to see all of the stuff contained deep inside your notes.

To see all of your attachments in notes, open the **Notes** app and tap the icon in the upper right hand corner that looks like a **circle with three dots** in it.

Now, hit the option to **View Attachments**.

You'll immediately see all of your Attachments, organized by type, along with little thumbnails of each.

To get a closer look, just tap one of the items and you'll get a larger view of it.

From here, you can use the share button in the lower left hand corner to share out or you can tap **Show in Note** in the upper right hand corner to bring up the note that contains this attachment.

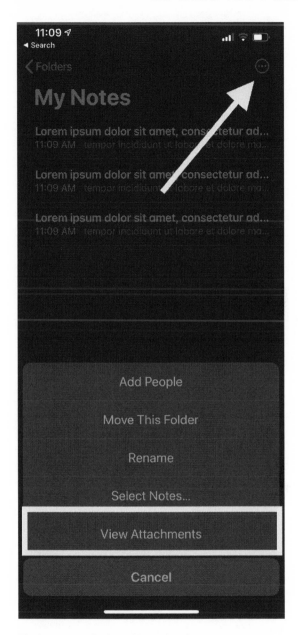

Tap here to see all of your Notes Attachments

Either way, this quick visual glance at your attachments can help you find things faster than ever. Or, it might be a trip down memory lane to see the stuff you attached ages ago and totally forgot about.

ONE OF THE most popular tips from the previous version of my book is how to scan a document using your iPhone. Apple has upgraded this feature to make it even easier to use, and it no longer relies on the Notes app.

To be fair, you can still use the Notes app to scan, store and even sign and share a document from, but there's also a new and easier way: the Files app.

To start, open the **Files** app on your iPhone. You'll see several **Locations**. If you don't, simply hit the Browse button in the lower right hand corner of your screen until you reach the main page with the word **Browse** at the top.

From here, tap the button in the upper right hand corner of the screen that looks like **three dots in a circle**.

This will reveal a menu where the top option is **Scan Documents**.

Tap it and your phone launches the camera and is ready to scan a document.

Just hold it over a piece of paper and it should find the edges and automatically scan. When you're finished, hit the **Save** button.

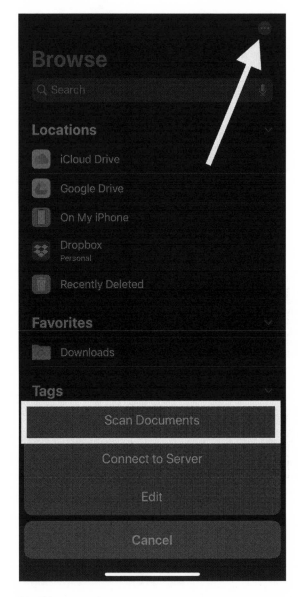

Use Files to scan documents

Now, you need to choose which folder to save it in. When you've selected one, hit Save in the upper right hand corner.

If you want to do more with your new PDF, navigate to the folder you just saved it in, find the file and tap it.

In the upper right hand corner, you'll see a button that looks like a **pen in a circle**. Tap here to mark up your document with a pen, pencil, highlighter and more. You might notice some new, finer controls including more precision over the thickness of the lines and more colors to choose from.

If you need to add **Text**, a **Signature**, arrow or shape to your page, just hit the **plus sign button** all the way towards the right.

Keep in mind if you add a signature, your iPhone can store it for the next time. It can even store several if you want!

When you're done making any changes, hit the markup button again (the pen in the circle) to exit edit mode.

You can now hit **Done** to exit out of your document, or hit the **Share** button in the lower left hand corner. This will let you AirDrop the file to a nearby device or send it off via email, Messages and more.

Keep in mind, you can now scan and save a document in any folder inside the Files app. You might even want to create a few folders where you can instantly scan items into. For example, a Receipts folder in iCloud would be a great solution if you travel a lot for work.

Now, remember what I said about how this feature used to be exclusive to the Notes app? Well, it's still there too.

To access a similar scan function, just open the **Notes** app and compose a new note.

Then, hit the **Camera** button above the keyboard and you'll see that the first option is to **Scan Documents**. Tap it to go through the similar process we just did in Files. The biggest difference is how you mark up your document in Notes.

Tap a scanned document inside a note to bring it full screen, then hit the **Share button** in the upper right hand corner to bring up the **Markup** option. It's a little more hidden, but you get the same tools to draw, sign and add shapes to your page.

Super double bonus new feature: When iPhone scans your document in Notes, it's also applying OCR, or optical character recognition, to the text on your pages.

This means that your iPhone can recognize the text that's actually on your page and you can search for a keyword to easily find your document later. In my testing, it's not perfect, but it's a neat feature that helps you find a page later if you can remember a word or two printed on it.

This is incredibly handy for receipts: just scan a receipt into Notes, and when you need to find it later, just type in the name of the merchant.

Like a lot of search bars, the one in Notes is generally hidden. Just open the app and pull down on your list of notes to reveal it.

53 / HAVE YOUR PHONE VIBRATE
WHEN FACE ID UNLOCKS IT

HERE's a small but useful feature that can save you some time when you're using Face ID to unlock your phone. Now, instead of looking at the little lock on the screen to see if Face ID was successful, you can set your iPhone so it makes a small vibration when Face ID unlocks it.

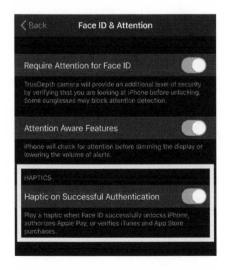

Your phone can vibrate when it unlocks using Face ID

It's a little thing, but once you turn it on, you might enjoy the difference because you can physically tell if your phone is unlocked just by feeling the confirmation vibration.

To set it up, head into **Settings** > **Accessibility** > **Face ID & Attention**.

You'll see an option for **Haptic on Successful Authentication**.

Turn it on and you'll feel a quick vibration when Face ID unlocks the phone, authorizes Apple Pay or verifies purchases.

EVERYTHING ELSE

I'VE SEEN IT ALL: some iPhone users give their devices elaborate names, others stick with the default iPhone or don't quite know how to change the name of their device.

Giving your iPhone a unique name can help out in various situations: identifying you for AirDrop, mobile hotspots and while connecting Bluetooth devices.

To set a name for your phone, go to **Settings** > **General** > **About**.

At the top is an option labeled **Name**, along with the current name of your iPhone.

Just tap here to give your phone a new name.

Keep in mind this isn't necessarily private - others might see it when your mobile hotspot is on, you're using AirDrop or making a Bluetooth connection.

So make it memorable but not something that gives away any deeply personal information.

When you're finished naming your phone, save your new name by pressing **About** in the upper left hand corner of the screen.

Tell your friends about this one: it will make AirDrop so much easier.

Give your iPhone a proper name

WiFi is a beautiful thing, but what happens once a WiFi network outstays its usefulness? Forget about it!

No, seriously, you can tell your iPhone to forget a WiFi network that you no longer want to use or need anymore.

This can be handy if you are connected to a network and it's the wrong one, the speed isn't as advertised or if there are two competing networks and you want your phone to settle on using just one of them.

Open **Settings** > **Wi-Fi**. You should see the network you're currently connected to along sections that might include **MY NETWORKS**, **PERSONAL HOTSPOTS** and **OTHER NETWORKS**.

MY NETWORKS is new for iOS 13, and it will show you any nearby networks that you've previously connected to .

Whether you're currently connected to a network or it's in the list of previous networks, you can forget a network and prevent your phone from connecting to it again.

If you want to "forget" a WiFi network you've connected to, hit the icon next to the name that looks like an **i** in a circle.

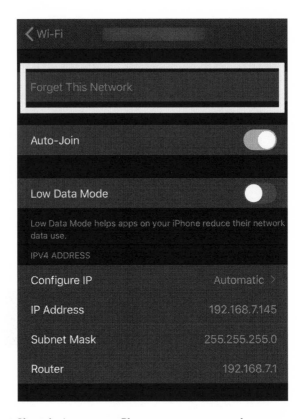

If you don't want your iPhone to connect to a saved
WiFi hotspot, tap Forget This Network

On the next screen, tap **Forget This Network**. Your iPhone
won't connect to this network anymore. Keep in mind, you won't see
an option to Forget in WiFi networks listed under **OTHER
NETWORKS** since you haven't connected to them before.

If completely forgetting a network is too extreme, you can toggle
the option for **Auto-Join**. If you turn this off, your iPhone will still
save the network login information so you can easily connect later,
but it won't automatically connect to the network. You will have to
specifically tap it in your list of networks for it to connect again.

This could be handy for a network that you might come into
range with often, but you want to decide when your phone actually
connects to it.

One more setting you might want to tweak while you're here in the Wi-Fi menu: **Ask to Join Networks**. Before iOS 13, there was just one option to toggle it on or off.

Now, there are several options including **Off**, **Notify** and **Ask**.

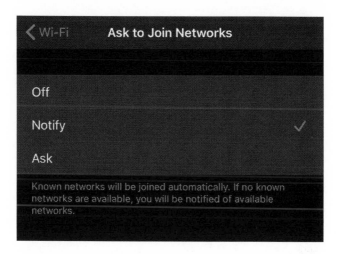

Off gets rid of any open network notifications all together. This is best for power users and anyone who would rather stick to their data plan over random WiFi hotspots. You can still connect to WiFi networks in settings.

Notify will present you with a list of available networks nearby so you can connect if you don't already have a WiFi connection on a known network.

Ask will attempt to find a good WiFi network for you using some Apple magic and wisdom of the crowds. You'll always have an option to say yes or no.

Should you use random WiFi networks? Generally, I'd say stick to known hotspots, like those supplied by a mall, restaurant or shop that you can identify.

Also, while you're on a public WiFi network I'd keep things like banking activities to a minimum. Your information is generally encrypted and kept private, but better to be safe than sorry.

EVER WONDER how long you have left on your iPhone warranty? There's an easy way to check.

Go into **Settings** > **General** > **About** and look for the section labeled **Warranty**. It might say **Limited Warranty** or **Apple-Care**, depending on whether you paid for an extended warranty.

You can see your expiration date here, or tap to see more details about your warranty coverage, including the estimated expiration date and links to more details about covered repairs.

This is a convenient feature to have, especially if your phone is acting up and you're waffling on whether to take it into the Apple store for a look. Seeing your warranty status can give you an idea of whether a repair might cost you out of pocket.

ARE your apps constantly bugging you to rate and review them in the App store? I understand why: developers depend on ratings to help gain traction in the App Store. Good ratings help drive downloads and purchases.

But, if you feel you can handle ratings and reviews on your own, without prodding from the developer, you can turn off these notifications.

To do it, just go into **Settings** > **iTunes & App Store**. Then, scroll down until you see the option for **In-App Ratings & Reviews**.

Toggle the switch to off and you won't be bothered again for feedback. If you find that you miss the little messages from time to time, you can always turn the toggle back on.

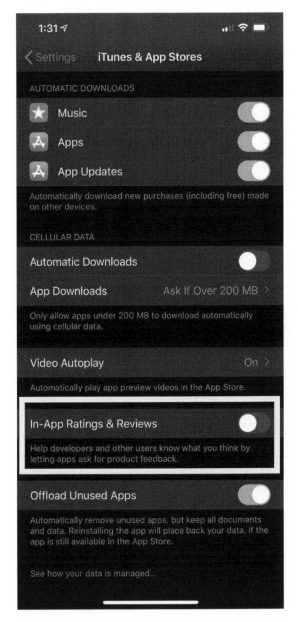

Turn off this setting if you don't want apps bugging you
for a review

IF YOU LOVE an app but don't think it deserves your cellular connection, you can tweak a setting so it can't use your data plan.

For instance, let's say you love your streaming video app but you know it gobbles up way too much data. For this reason, you only want to binge watch your shows when you're on a WiFi connection.

Instead of checking your connection every time before you stream, you can just change a setting and the app won't be able to use your data.

To do it, go to **Settings** > **Cellular** and look for the section labeled **CELLULAR DATA**.

Depending on how many apps you have installed, it might take a minute or two for the entire list to show up.

You should see some information about how much cellular data you've used in total, along with each individual app and how much data it has used recently.

Next to each app is a toggle switch. Just find the app that you want to cut off cellular data to and flip the switch. That's all there is to it - the next time you open that app and you're on a cellular connection, you might get a warning that the app can't access data.

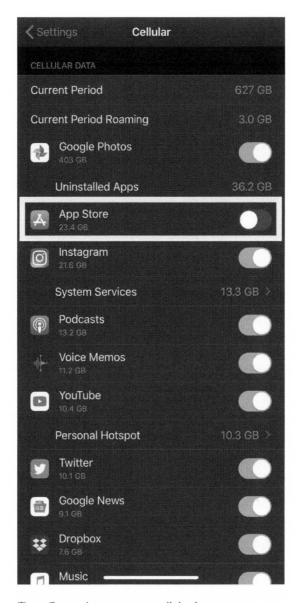

Turn off an app's access to your cellular data

The app itself won't refresh until you reinstate access to your data plan or you're on a WiFi connection.

It's an easy way to manage access to your cellular data, especially if you're on a limited plan.

59 / SEE WHICH APPS ARE STORAGE
   HOGS

IF YOU NEED to free up some storage on your phone fast, it might be easiest to see which apps are taking up the most storage. This way, you can delete the app completely or go in and clear out some of its data.

Believe me, I can't tell you how many school plays and birthday parties I've been at where I witnessed that dreadful out of storage message on another parent or grandparent's phone as they try to snap a picture of their little one.

To see which apps are taking up the most storage on your phone, go to **Settings** > **General** > **iPhone Storage**.

In here, you will see a breakdown of your iPhone storage along with a list of app sorted by how much storage they're taking up.

Tap an app to see an even more precise breakdown of the storage it's using. There are also options to **Offload App**, **Delete App** or there could be additional options to delete some of the individual files taking up storage inside the app.

The quickest and easiest thing to do to free up storage is to identify an app that you don't really need taking up a lot of data. You can simply select the app and hit **Delete App**.

See which apps are taking up the most storage on your phone

If it's something more nuanced, like a video app, you might see all of the **DOWNLOADED VIDEOS** that app is currently storing. You can swipe right to left on a video or file to delete just that item.

No matter how you proceed, this can be a useful tool for some storage triage on your phone. Just be careful that you don't delete something that isn't backed up or can't be restored if you need it again.

IF YOU'VE CHOSEN some favorite sports teams in the TV app on your iPhone, iPad or Apple TV, you might see tune in notifications from time to time. These remind you of upcoming games.

But, your iPhone (and perhaps Apple Watch) could spoil those games all together by showing sports scores as notifications before you're ready to see them.

To keep this from happening, there is an option to turn them off.

To change it, go into **Settings** > **TV** and look for the option that says **Show Sports Scores**. Don't want to see them? Toggle the switch off.

This way, your phone might show tune in prompts, but it will not show you the score of the game.

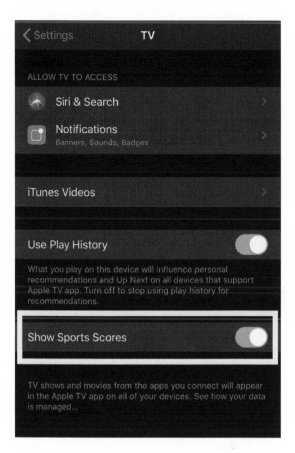

Make sure Sports Scores spoilers don't ruin the game

DID you know your iPhone can say the name of the person who's calling? This is handy when your phone is across the room or in your bag and you want to know who's calling without looking.

To set it up, just head into **Settings** > **Phone** > **Announce Calls**.

You'll have several options to choose from, including **Always**, **Headphones & Car**, **Headphones Only** and **Never**.

Pick the one that's right for you: maybe you always want your phone to tell you who's calling or just when you have headphones in.

Whichever option you choose, hearing the name of the caller can be a helpful addition to your next incoming phone call.

Keep in mind, you'll only hear a person's name if they are in your address book. Otherwise, you'll hear Unknown Caller.

This doesn't necessarily mean it's a robocall or there's no Caller ID info, if you look at your phone you will still see a phone number.

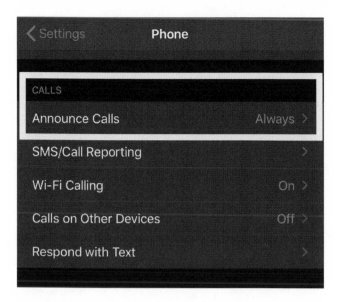

With Announce Calls on, you'll always know who's calling

WE ALL KNOW it can be tricky to get the perfect shot, especially when trying to take a selfie.

Next time, instead of trying to balance your iPhone in one hand as you struggle to reach the shutter button with your thumb, just use a **Volume key** to take your picture instead.

Either one will work, so get your phone in a comfortable grip, duck those lips and say "cheese!"

You'll find that it's much easier to press the volume up or down key to activate the shutter, especially when your arm is stretched out far in front of you.

And before you tell me you hate selfies, keep in mind this tip works for regular photos too.

EVER TAKE a picture that you would rather no one else see?

For instance, I once took a picture of myself wearing one of those face masks my wife loves to put on and instantly regretted it. Both the picture and putting on the mask.

In the end, I liked the way my skin felt but still hated the picture.

And, although I didn't want to post it to social media, I wasn't ready to delete it. So I hid it.

This means that I can still bring up the picture when I want to embarrass myself, but if I hand my phone to someone and they scroll through my recent pics, they won't be subjected to the nightmare imagery.

I know, harsh example. Here's how to do it.

Open **Photos** and find the offending picture. Tap the **share icon** and scroll down until you see the option for **Hide**.

You'll get a confirmation message explaining that the picture will be hidden from all places in your iPhone photos library but it can still be found in the **Hidden** album.

After you hide a picture, if you want to see it, go to the main page of the Photos app - the one that has a bunch of sections at the bottom including **Photos**, **For You**, **Albums** and **Search**.

Hide a photo so it doesn't show up in your Camera Roll

Tap **Albums** and scroll all the way down until you see a section labeled **Other Albums**. In here, you can tap the **Hidden** album to see your hidden pictures.

Ironically, they are just one step above **Recently Deleted**. You have to love Apple's sense of humor.

By the way, if you ever deem a hidden picture fit enough to return to the main photos area, you can tap the share icon once again and choose **Unhide**. It will return to its original place among the regulars.

WHEN YOU DELETE a photo on the iPhone, it doesn't really go away instantly. It is actually stored for about a month in a special album called **Recently Deleted**.

This can be useful if you delete a photo by accident and want to retrieve it. Alternatively, it would be the first place I'd look for incriminating evidence if I ever picked up someone's phone.

Just kidding.

Either way, it's good to know that it exists and understand how it works.

It can also be handy for freeing up space on your phone fast if you have a few big videos in there.

To find it, open the **Photos** app and tap the section at the bottom labeled **Albums**.

From here, scroll all the way down until you see the section labeled **Other Albums**. Tap **Recently Deleted** to see the photos and videos you've recently deleted.

You'll notice each photo is marked with time stamp, this is how long it will remain on your phone until iOS automatically deletes it permanently.

They're usually deleted in 29 days.

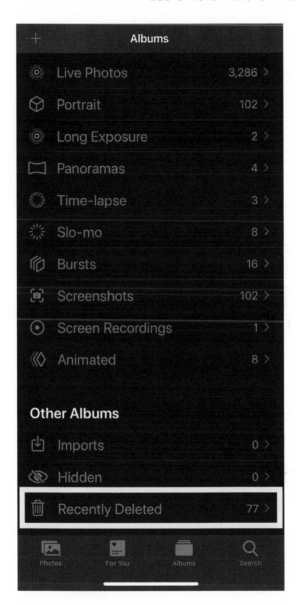

Scroll all the way down in your Albums to see Recently
Deleted photos

Tap on a photo to bring up two options: **Delete** and
Recover.

These are pretty self-explanatory. Tap **Delete** to eradicate that photo from your phone instantly. Tap **Recover** to put it back into your main Camera Roll.

Back on the main **Recently Deleted** screen, you'll notice a **Select** button in the upper right hand corner.

Tap it to bring up two new options: **Delete All** and **Recover All**.

They both do just what you think. Get rid of everything instantly or restore all of your pictures instantly.

However, there is another option once you tap **Select**: you can now pick and choose several photos at a time to **Delete** or **Recover**.

Whether you perform these tasks with a few pictures or all at once, you'll always get a confirmation button that details how many photos you are Deleting or Restoring as sort of one final reminder of your actions.

CREATING photo albums used to be a national past time, right up there with baseball. But these days, we're sharing to social media in real time and no one seems to have time to create an entire album anymore.

Still, you are free to create your own albums on the iPhone, but it also creates some albums for you. To see them, open the **Photos** app and tap the section at the bottom labeled **Albums**.

Under the section labeled **My Albums**, you'll see an important one: **Favorites**. This is a useful album because it contains the pictures you've marked with a "heart."

These can be the pictures you love, the ones you plan to post to social media later or just assorted snapshots that contain information you might want to refer back to from time to time.

To mark a picture as a favorite, go back to the **Photos** tab, tap to choose a photo and then tap the **heart** in the bottom center of the screen.

Now, go back into the **Favorites** album and you will see the photo you just "hearted" there. This makes finding pictures later much easier than scrolling through a sea of snapshots.

	Albums	
Media Types		
☐ Videos		523 >
☐ Selfies		554 >
◎ Live Photos		3,286 >
◇ Portrait		102 >
◎ Long Exposure		2 >
☐ Panoramas		4 >
✾ Time-lapse		3 >
✦ Slo-mo		8 >
⦿ Bursts		16 >
⦿ Screenshots		102 >
⦿ Screen Recordings		1 >
⟨⟩ Animated		8 >
Other Albums		

See all the types of pictures you've taken, including Selfies

The other interesting albums here are listed under **Media Types**.

These are automatically created by your iPhone and contain pictures of the same kind: **Videos**, **Selfies**, **Portraits** and more.

If you're looking for **Screenshots**, **Panoramas**, **Slo-mo** or another type of picture or video you remember taking, this might make it easier to find.

Admit it: you want to see all of your selfies; now you know where to find them.

NAMING the people in your pictures makes it really easy to find all of your pictures containing them.

You can also create a quick little movie containing photos of them instantly - which is handy for sharing on social media on their birthday.

To do it, open the **Photos** app and find a picture with a friend or family member in it. Tap to see the single picture full screen and then swipe up on it.

This will bring up additional details about the picture including a section labeled **People**. It should show the faces of anyone inside the pictures.

Tap a face and at the top you'll now see an option to **Add Name**.

As you start to type their name, if the person is in your address book, the name will auto-complete with this information. You might also be asked to confirm additional photos containing this person's face.

Once you've named a person and there are enough pictures of them in your library, you'll be able to create a little movie of them from your photos.

Swipe up on a photo to assign names to the people in it

To do this, go into the **Albums** section of the Photos app and scroll down until you see the section labeled **People & Places**.

Tap **People** to bring up all of the people you've identified in your photos.

Tap one of the people and hit the play button. iPhone will create a movie instantly from these pictures. You can watch and share as is, or hit the **Edit** button in the upper right hand corner to change the music, length or even the photos contained inside.

If you don't see the movie option for someone, that means you need to take some more pictures of them so there's enough media to create one.

WHEN YOU SIGN in on your iPhone with all of your various accounts, they are usually added to your calendar by default unless you turn them off.

This can result in multiple calendars displaying duplicate information, or calendars showing information you don't need or use.

To fix this, you can hide or unhide the calendars you don't necessarily want there.

To do this, open the **Calendar** app and tap **Calendars** at the bottom of the app. This will bring up a list of all of the calendars available to you on your phone. The ones with checkmarks next to them are currently being displayed when you browse the calendar.

To hide the calendars you don't need or want shown, just tap the checkmark next to one of them. On the flip side, if a calendar isn't selected and you want it shown, tap the circle to select it.

Want to change the color of the events shown in a particular calendar? Just hit the icon that looks like an **i** in a circle next to the calendar. You can choose a color here so you can instantly identify the events from that particular calendar.

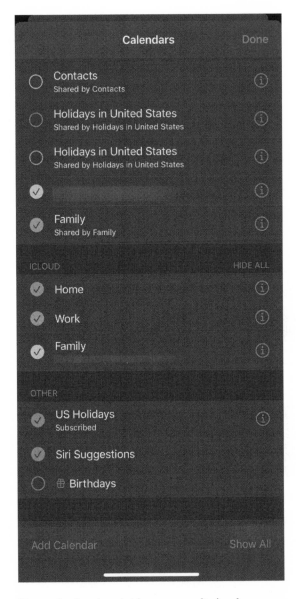

Tap to select just the calendars you want displayed

From inside the **Edit Calendar** screen you can also delete a calendar completely. Just scroll all the way down to the bottom and look for the **Delete Calendar** button.

There's one more calendar tip to know: it's a good idea to choose a default calendar. This way the events you create will always start in the correct calendar - especially if you use Siri to create them.

To confirm or set your default calendar for events, go into **Settings** > **Calendar** and look for the option for **Default Calendar**. It could be set to an iCloud calendar when you really want it set to your Google, Microsoft or work account.

Just tap and you can choose from all of the calendars available on your linked accounts.

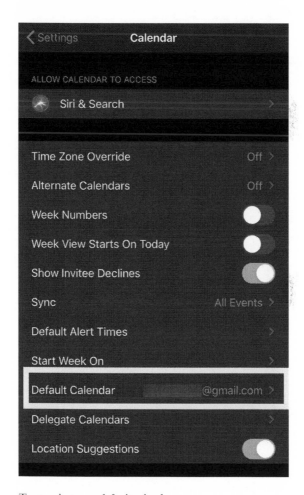

Tap to select your default calendar

FROM TIME TO TIME, I get questions from folks about items disappearing out of their calendars. Usually, it's when they're searching for a past event they know was in their calendar, and it doesn't show up.

The culprit: a setting that tells your iPhone how far back to sync your calendar events.

Personally, I like all of my events - past, present and future - at hand for easy reference. You can choose what works for you.

To set it, go into **Settings** > **Calendar** > **Sync**.

Here, you have various options ranging from **2 Weeks back** to **All Events**. Pick the length of time that works for you.

This way, you can easily scroll or search for entries in your calendar, even if it happened a while ago.

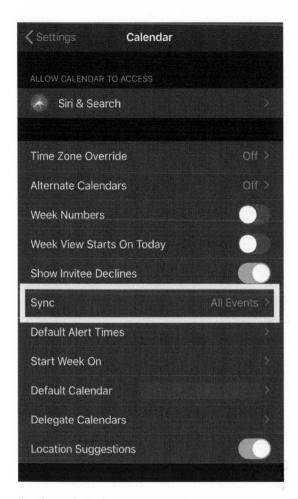

Decide how far back you want your calendar to sync

THE NOTES APP gets a big upgrade in iOS 13, but there is one secret feature that remains the same: the ability to "pin" a note to the top of the list.

This can come in handy when you want to keep a particular note above all others for quick reference.

It's so hidden in plain sight that you might never discover it on your own.

To pin a note, open the **Notes** app and go to a list containing your notes. Then, just **swipe from left to right** on the note you want to pin.

This will reveal a **pushpin icon** to the left of your note.

Tap the icon and your note will stay at the top of the list.

Alternatively, if you do a slow swipe on a note from left to right, it will reveal the push pin icon and spring back - but when it springs back your note will now be pinned.

Try both ways and see which one is easier for you.

Now, no matter how many notes you save, your pinned note will always show up at the top of the list under a header labeled, you guessed it, **PINNED**.

Swipe left to right on a note to pin it to the top of the list

Want to unpin a note? Just try the two methods in reverse.

Pull a note in the list to the right to reveal a new pin icon, one that is now crossed out. Tap it to unpin the note.

Or, you can quickly pull a note to the right to unpin it automatically.

No matter which method you choose to pin and unpin, it sure beats sticking yourself with a sharp pin by accident in real life.

FREE TRIALS for the Apps you download can be a tricky thing.

You want to sign up to try something out, but you know you're inevitably going to forget to cancel before the free trial is up. Next thing you know you've been charged for a subscription.

Here's a better way: cancel the trial as soon as you sign up.

This way, you don't forget to cancel and your account isn't charged, but you still get the full use out of your trial. If you like it, you can just re-subscribe - on your terms.

At first thought, this might seem like a sneaky or dishonest thing to do. You're joining a service with the intention of never actually signing up for it. And sure, that might be true, but I take a more optimistic approach: you're joining a service and getting to try it out on more relaxed terms.

By not getting charged for a service you forgot about, you don't start off a relationship with the app or service on the wrong foot.

Nine times out of ten, you can cancel the subscription immediately and still have full access to the free trial time period. The only subscription services I've seen that this trick doesn't work for are some of Apple's own, like News+.

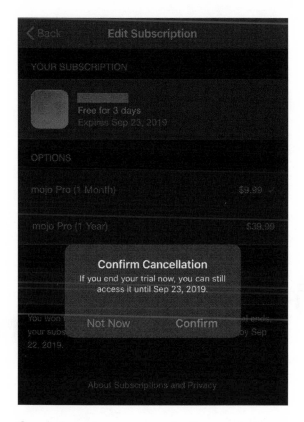

Cancel a free trial as soon as you subscribe so you don't forget about it

If you cancel before the free trial is up, you'll lose access to the free trial immediately.

Here's how to do it.

First, sign up for a subscription. This could be a free trial of any app or service in the app store.

Before you confirm your subscription through the App Store, be sure to read the POLICY and PRICE sections closely. It should say something like "cancel anytime in Settings at least a day before each renewal date." Under price, it should state the length of the free trial and the date you would normally get billed for your first payment.

As long as it's actually a free trial, you'll get the full use out of the time before you get billed.

Confirm the subscription, then close out the app and head into **Settings > Apple ID, iCloud, iTunes & App Store**. This is the first option in settings above Airplane Mode.

Next, tap the section labeled **Subscriptions**.

You will see all of your current subscriptions listed under **ACTIVE**. Locate the service you just subscribed to and tap it.

Then, hit the button labeled **Cancel Subscription**.

You should see a confirmation box that says you can still access your free trial until the original expiration date. Hit **Confirm**, and the subscription is cancelled.

Now you can use the app as normal with the free trial without worrying about getting charged for something you don't want.

If you love it, just subscribe for real next time. I find that this little trick takes so much pressure off of the free trial process that eventually it works out in the App's favor.

Because you're less worried about being charged accidentally, you actually explore more paid app subscriptions. Eventually you'll find one that you like, and stick with it, which is a good thing for the developer.

I NEVER REALLY UNDERSTOOD THIS one, but for some reason, used boarding passes don't ever go away on an iPhone.

Perhaps someone at Apple likes the idea of keeping them around for posterity, but when I'm done with a flight, I don't need my boarding pass lingering around for ever and ever.

Yes, I'm talking about the **Wallet** app. Let me show you how to clean this thing up. This assumes that you have saved a boarding pass to your phone at some point.

First, open the **Wallet** app. You will see any payment cards you've linked to Apple Pay at the top.

Scroll down if you need to and look for an old boarding pass. Tap it to bring it center on the screen. It should say "this pass has expired" with a greyed out QR code.

Once you've confirmed that the pass is no longer needed, hit the button in the upper right hand corner that looks like it has **three little dots** in it.

This used to "flip over the card," but now it just brings you to the settings screen.

Towards the center of the screen, you should see an option to **Remove Pass**.

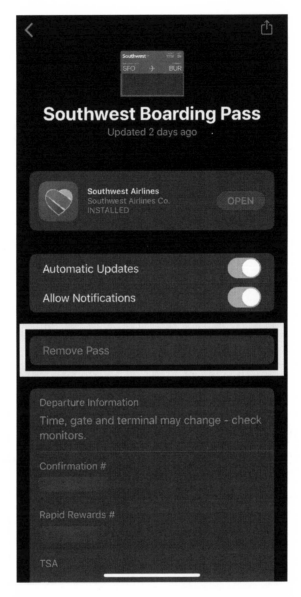

Delete a used boarding pass

Tap here, confirm the removal and watch the pass disappear.

I'll leave you some time to go organize your Wallet. I'll be right here when you need me again.

GOT a number in your recent call list that doesn't belong there? Maybe it was a misdial, spam or a friend who you no longer talk to? Just kidding, that last one would be harsh.

Whatever the reason, if you need to delete a number or few out of your Recents list, there's a fast and easy way to do it.

First, open your **Phone** app (you know, the one you use to place calls) and tap **Recents** at the bottom. Next, find the offending number and simply **swipe left to right** on it to reveal a **Delete** button shrouded in red.

Tap **Delete** and the number disappears.

Want to get rid of a bunch of numbers at once? Just hit the **Edit** button in the upper right hand corner of the screen.

This will reveal a **Clear** button in the upper left hand corner of your screen. Hit it to reveal yet another button labeled **Clear All Recents** at the bottom of your screen. Get the feeling Apple doesn't want you to do this?

When you're ready to proceed, hit the **Clear All Recents** and watch them go away.

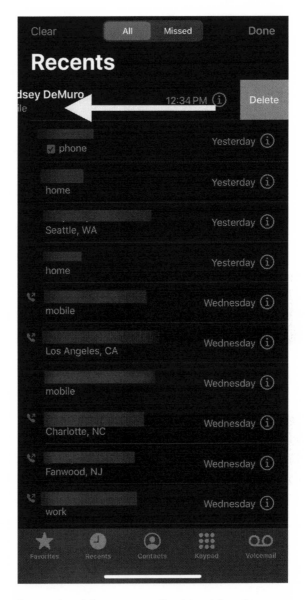

Swipe right to left on a recent call to delete it from the list

WHILE WE'RE in the **Phone** app, let me show you how to find the secret redial key. You've probably never seen it because it's not labeled, but you've hit it many, many times.

First, tap the section at the bottom of the screen labeled **Keypad**.

Now, see that **big green button** that you use to dial a call? It also doubles as a Redial key. Don't believe me?

Without dialing any digits, tap it once.

Did the last number you dialed appear above the dial pad? Pretty cool, right?

This came in handy for me when I was disconnected from a call and needed to redial quick. I just tapped the green call button and was on my merry little way.

It might not be life changing, but it sure is fun to know about these things.

The call button doubles as a redial key

PODCASTS ARE ALL the rage these days. Although they have been around for some time, there is a renewed interest in the audio programs and that's fine with me.

Part of their appeal is that you can find podcasts on just about any topic you're interested in and you can listen on your schedule.

Although the idea of podcasts can be intimidating, they don't have to be. I get that radio is always on and generally starts up with your car, but your favorite podcast can be just as easy to listen to thanks to the help of Siri.

You can ask Siri to play just about any podcast out there with a simple voice command.

My podcast is called **Rich on Tech**. In it, I talk about the tech stuff I think you should know about (much like this book) and answer the questions that you send me. We'll use mine as the example here.

Let's say you hear me say to "check out my podcast called Rich on Tech."

First, grab your phone and activate Siri. If your phone has no home button, just press and hold the side button until she appears on screen. If you have a home button, just press and hold it.

Now, just say "**Listen to the Rich on Tech Podcast**."

Just ask Siri to listen to a podcast

Immediately, it will start playing.

If you have "Hey, Siri" turned on (**Settings** > **Siri & Search** > **Listen for "Hey Siri"**) you don't even have to touch your phone to play a podcast. Just say "**Hey, Siri, play the Rich on Tech Podcast**" and she will immediately start playing it for you.

In the car? You can do the same thing. If you have CarPlay or your phone is connected via Bluetooth, just use the same "Hey, Siri, play the Rich on Tech Podcast" and it will begin playing.

Also, if you can get fancy with some qualifiers. For instance, I sometimes ask to listen to "**the latest episode**" of a podcast. Try and see what else works!

Alternatively, you can press and hold the on-screen home button (which now looks like a grid of buttons) to activate Siri and use the same commands above to play your show.

One more thing: please don't forget to rate and review my podcast. Thanks!

IT PAINS me to write this tip, since I know you'll probably use it to skip the commercials in your favorite podcasts, but the flip side is that if you don't adjust this podcast setting, it can lead to some frustration in the car.

Let's say you're listening to a podcast using Apple's Podcasts app in your car or with headphones. You get to a boring part that you would rather skip and so you hit the skip forward button on your car's steering wheel controls or on the headphones.

By default, this will take you to the next podcast in your queue, which is probably not what you probably wanted to do.

More likely, you just wanted to skip ahead 30 seconds or so.

The good news is that Apple offers a way to adjust how these external controls work. By default, if you press them, they will skip to the next or previous episode. But here's how to change the setting so you can just move forward a bit in your current show.

Go to **Settings** > **Podcasts** and scroll all the way down to the section labeled **EXTERNAL CONTROLS**.

If you haven't changed anything, there is likely a checkmark to the right of **Next/Previous**.

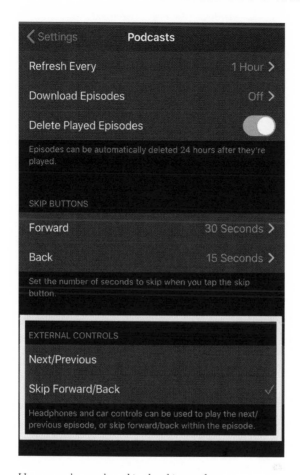

Use external controls to skip ahead in a podcast

To change the behavior to **Skip Forward/Back**, just tap it and the checkmark will move, signifying your new default.

Since you're already here, you might also want to adjust how much these buttons will skip forward or back.

Look in the section labeled **SKIP BUTTONS** to see the current values and tap to adjust to your liking.

You can skip forward anywhere or back from 10 to 60 Seconds at a time. Personally, I think 30 seconds forward at a time is just enough, while 10 or 15 seconds back is good in case you skip ahead too far.

Back out of all these setting screens when you're finished and you will now have much more control over your podcast playback.

WANT to instantly recognize when your partner, spouse, parents or kid is calling calling you? Set a specific ringtone for them!

It's a fast and easy way to identify important callers without having to look at your phone.

Alternatively, you can avoid those calls that you might not want to waste your time running across the room to answer.

To do it, open the **Contacts** app and search for the contact you want to customize.

Next, tap their name to bring up their card and then hit the **Edit** button in the upper right hand corner of the screen.

Scroll down a bit until you see the section labeled **Ringtone**. It is probably set to **Default** by, well, default.

But you're going to change that!

Tap the **Ringtone** area to bring up your options, listed under **RINGTONES**.

Tap the name of a ringtone to get a little preview of what it sounds like. When you're happy with your selection, hit **Done**.

Now, the next time that person calls, you'll know exactly who it is just from their ringtone. Assuming you remember which ringtone you assigned them.

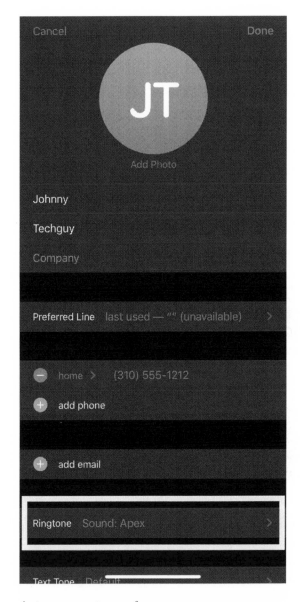

Assign a custom ring tone for a contact

WHILE YOU WERE SETTING that custom ringtone, you might have noticed another useful feature inside the ringtone selection screen called **Emergency Bypass**.

This will make you feel better when using the Do Not Disturb feature, which silences incoming calls and notifications when it's activated.

Some people are hesitant to enable it because they fear they might miss a call. I'm always surprised when I ask people if they put their phone on Do Not Disturb when they're sleeping at night and they tell me "No" because they don't want to miss an important call.

With **Emergency Bypass** enabled for a contact, your phone will still notify you even if you're in Do Not Disturb mode.

To turn it on for a contact, go into **Contacts**, tap the person's name to bring up their card, hit **Edit**, then tap where it says **Ringtone**.

Now, toggle the switch next to **Emergency Bypass**.

The next time they call while your phone is in Do Not Disturb, your phone will ring as usual.

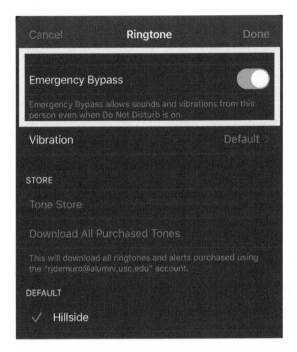

Emergency Bypass ensures important contacts can
always reach you

Keep in mind, there is another way to allow incoming calls during
Do Not Disturb. Just go into **Settings** > **Do Not Disturb** look for
the option to **Allow Calls From**. You can select from **Everyone**
(which would sort of defeat the purpose of the setting), **No One**,
Favorites or choose a group you've set up.

Mine is set to **Favorites** - if these people call, they will still ring
my phone even if it's set to Do Not Disturb. I can rest easy and still
get important calls if necessary.

I HOPE you never lose your phone, but if you do, it's a good idea to check these settings before the fact.

They can help you find your phone if it's ever lost or stolen. But, if they're not turned on, you won't be able to use these helpful features.

For this reason, now is a good time to make sure these features are turned on. Usually, they should be activated during the setup process, but let's make sure.

To start, go into **Settings** > **Apple ID, iCloud, iTunes & App Store** > **Find My**. From here, look at the top option for **Find My iPhone** and be sure it's toggled **On**. If it's not, tap it to bring up the option to turn it on.

Once you do, you get two more helpful tools you'll want to turn on. The first, **Enable Offline Finding**, is new to iOS 13 and potentially super helpful.

It sort of turns your iPhone into a Tile. You know, those little Bluetooth devices that can help you locate your keys, wallet or other things you attach them to.

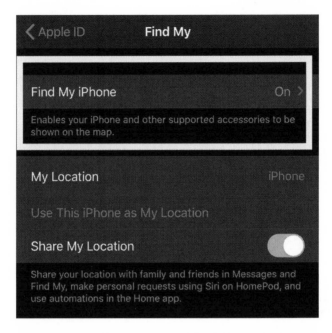

Make sure that Find My iPhone is turned on before you need it

Apple knows that a lot of times, there might not be an active internet connection on your lost device so it is taking things a step further.

Offline Finding will use the Bluetooth on your device to essentially send out a little distress signal to nearby Apple devices. They recognize the signal and pass it on to your account to help you locate your lost device.

The final option to enable is **Send Last Location**. Turn this on and your phone will send its final GPS coordinates before the battery completely dies. This can be handy if you drop your phone in the woods and don't find it for a few days.

Even though the battery has gone dead, you can still walk to the location where you dropped it.

That is, unless some sort of small animal picked it up and scurried it away to its underground bunker. In that case, you're out of luck.

But at least you have a cool story!

You're probably well aware of how the web browser on your iPhone keeps a record of every website you visit.

If you want to get rid of this browsing history, you can do this in Safari's settings.

Go to **Settings** > **Safari** and scroll down until you see the option for **Clear History and Website Data**.

Tap it to clear your history, cookies and other browsing data. Keep in mind, this will also clear your data on any other Apple devices logged into your iCloud account.

It will not affect your bookmarks, but you will have to re-login to websites again.

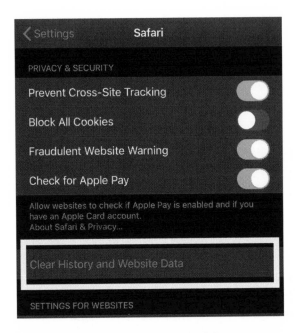

Clear your web browsing history stored in Safari

IN iOS 13, Apple is putting tools to change the look and feel of the websites you visit in a much easier to access place in Safari.

You can increase or decrease text size, hide Safari's toolbar, request a desktop version of a website and more - all right from a web page.

To access the tools, open **Safari** and navigate to a website, perhaps a news site.

Bring up an article and hit the button that looks like **two letter AA's** in the upper left hand corner.

Now, you'll see various tools.

The first will increase or decrease the text size. Hit the **bigger A** to increase the font size, hit the **little A** to decrease the font size. Tap the percentage number in the middle to bring things back to 100%.

Next is **Show Reader View**. Tap it to generate a clean version of the article you're trying to read, minus any extra items on the page like "ads or distractions."

When you're in Reader view, you can further change the look and feel of the page by tapping the two letter AA's in the upper left hand corner.

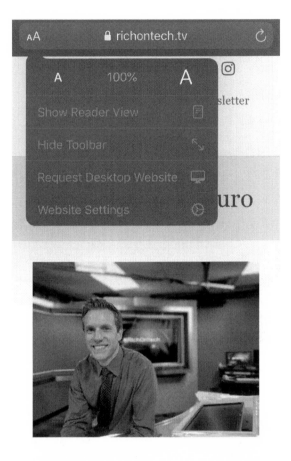

Rich DeMuro is the tech reporter
KTLA-TV Channel 5 in Los Angel
and appears on the #1 rated KTLA

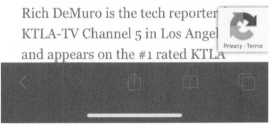

Tap the double AA's to access tools to make a website
easier to read

You'll see various fonts and background colors to choose from.
My personal favorites here are Georgia and Sepia tone.

Exit Reader view and the next option is to **Hide Toolbar**. This will reclaim some space at the top of your page usually occupied by the URL address bar, or the place where it shows the website address. To get the Toolbar back, just tap on the web address.

Other options inside the AA menu include the ability to **Request Desktop Website** - this is handy if the mobile version isn't doing what you want it to.

You can also **Turn off Content Blockers** if installed. These are apps that prohibit some ads from displaying on your web pages in Safari. Turning them off can help when a website isn't acting like it's supposed to. Sometimes, Content Blockers are the culprit.

The final option is **Website Settings**. Here, you can tweak the individual settings for this specific website.

Pro tip: if you want to skip the additional menus and just want a Reader View of a website, long press the double AA's. This will turn Reader View on and off instantly.

EARLIER IN THE BOOK, I told you about the new search tool available in Messages. There is a another way to find something you exchanged with someone in your texts. This might be an easier place to start if you remember the person you exchanged the information with.

This tool got an upgrade in iOS 13 to make things more organized and easier to locate.

Links, locations and attachments are now grouped together and separated by section so you can easily find what you need.

Let's say you you remember that your friend Sam sent you a link to a cool website that lets you see which bands are playing in your neighborhood. Instead of scouring through all of your texts with Sam, here's an easier way.

First, open the **Messages** app and tap on the conversation between you and Sam.

Once you're in, tap Sam's **profile picture** and/or phone number at the top. This will reveal a few new options including **audio**, **FaceTime** and **info**.

Tap the button for **info** to bring up a page labeled **Details**.

Tap info to see links, photos and more you've exchanged with
someone in Messages

Scroll down and you'll see various sections containing the items
you've exchanged including **PHOTOS**, **LINKS** and more.

Expand a section to see everything contained in it by tapping the
appropriate "**See All**."

In our example, we'll tap where it says **See All Links** and this
will bring up all the links we've exchanged, along with a helpful
thumbnail of the website.

Tap a link to open it up, or tap and hold on an item to bring up
more options, including a **Copy** and **Share** command.

Between the new and improved **Messages** search and the new
and improved **Details** pane, hopefully you won't have too much
trouble finding exactly what you need.

THERE ARE lots of little surprises hidden in Apple's iOS operating system. One of them is how to find the timestamp on a text message.

Usually, messages are time stamped every so often. I'm sure there is rhyme and reason to it, but I don't have the time to figure it out.

All I know is that sometimes you want to see the exact time a message was sent or received. Here's how to find that information.

Open the **Messages** app and go into a conversation. You'll probably see time stamps every so often in the center of the screen throughout the conversation.

To see the time stamp for a specific message, just put your finger on the center of the screen and **pull to the left**. This will reveal exact time stamps along the right side of the screen.

This could be handy for checking the time your kid said they would be home compared to when they actually arrived.

My kids are still too young for that reality, but when it's time, I will be ready. Oh, I'll be ready all right.

Swipe right to left on a message to reveal its time stamp

You MIGHT NOTICE that when you get a text or iMessage, your phone will remind you one more time if you don't take a look at the message.

To some, this can be annoying, but others might find it useful.

By default, iPhone will repeat a message alert once after two minutes, but if you don't find this useful, you can turn it off completely. Alternatively, if you want more repeat alerts, you can set that as well.

To modify this setting, go into **Settings** > **Notifications** > **Messages** and look all the way at the bottom for the section under **OPTIONS** labeled **Repeat Alerts**.

The default is likely set to **Once**, but you can tap to see all of your available options, ranging from **Never** to **10 Times**.

Whatever you choose, alerts will repeat at two minute intervals. So, if you chose the maximum 10 times, you could have your phone chirping at you to remind you to check your incoming message for about 20 minutes.

That's not terribly long, but look around and you'll notice that many people can't stand to look away from their phone for more than a few minutes.

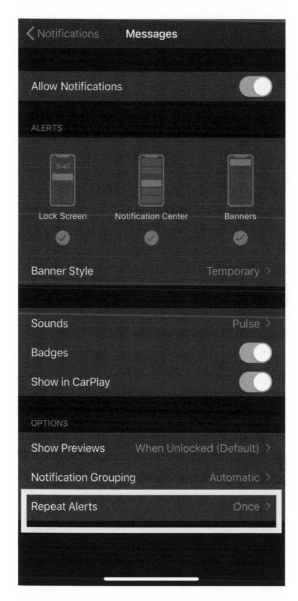

Decide how many times you want a message notification
to repeat

As far as I can tell, Messages seems to be the only pre-installed
app that lets you set a repeat for alerts.

ONE OF MY all-time favorite features on the iPhone is a pretty simple one: search.

Back in the day, Apple called the search function on the iPhone Spotlight, an homage to search on Mac computers, but these days it's often just referred to as Search.

There are several ways to access search on the iPhone.

If your screen is locked, tap to wake your phone then put your finger in the middle of the screen and swipe right. This takes you a Widget screen with a Search bar at the top.

Tap in here and you can begin your search. If your phone is still locked, you'll likely have to unlock it before you can get any search results.

The second way to access the Search bar is very similar to the first, except your phone is unlocked first. From the home screen, place your finger in the middle of the screen and swipe right. This will bring up the same widgets screen with the Search bar at the top. Tap to type in your search.

The third (and my favorite) way to access Search on the iPhone is to start with your phone unlocked and on the home screen. Now, just place your finger in the middle of the screen and pull down.

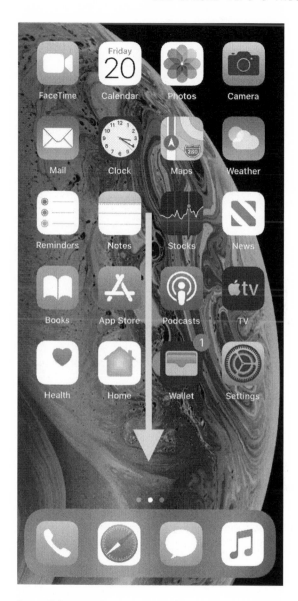

Just pull down on the screen to search

This will reveal the Search bar along with a results page. You can start typing in your search and begin to see results populate immediately.

Don't want to type? Just hit the microphone icon on the right side of the search bar and use your voice.

So, what is so great about the search bar?

Personally, I love it for two things: the ability to find apps and contacts immediately. For apps that aren't on your home screen, it's often easier and faster to just type in a letter or two of their name and watch them appear in the top section labeled **APPLICATIONS**.

My other favorite use is contacts. Just type in the name of the person you want to call, text, FaceTime or email and there they are. You don't have to open the **Contacts** app or even the **Phone** app.

The best part: Siri looks inside your emails for Contact information. So even if a person isn't in your address book, you can still find them if you've ever exchanged a message with them that contained some information like their name, email or phone number.

It's pretty magical, and one of the reasons why I recommend putting your Name, email address and phone number in your email signature. It makes it easy for Siri to index your information!

I CAN'T SAY for sure that having your emergency medical information stored and accessible on your iPhone can save your life or even help you out in an extraordinary situation, but I don't want to be the one to find out the hard way.

iPhone has a built in feature that lets others access vital medical information about you and call emergency contacts that you designate - and all without unlocking your phone.

To set it up, open the **Health** app, then tap your profile picture in the upper right hand corner. If you haven't set up your pic, it might just be your initials or a silhouette.

Under the section labeled **Medical Details**, tap **Medical ID**.

Proceed past the introductory screen to enter your details. You can be as specific or as general as you like. Keep in mind that people will be able to access this without unlocking your phone.

Tap to enter your **Date of Birth, Medical Conditions, Medical Notes, Allergies, Medications, blood type** and more.

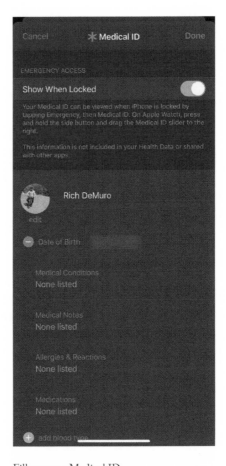

Fill out your Medical ID

Scroll down to the bottom of the screen to see a section labeled **EMERGENCY CONTACTS**. You can tap here to add people from your address book that your iPhone will be able to call even when it's still locked.

Before you exit out of this **Medical ID** screen, make sure that you have the **Show When Locked** toggle turned on (green). This way first responders or anyone else that needs to can access this information.

How, you ask? Let's find out.

Close out of the Medical ID screen and then lock your phone.

Now, double tap your screen or hit any button to light it up but do not unlock it. If you have Face ID, just hold your phone in a way that you can see the lock screen but it doesn't unlock.

Now, swipe up from the bottom of the screen to make your phone try Face ID (again, don't unlock it). When Face ID fails, you should see a screen where you can enter in your Passcode.

In the lower left hand corner of the screen, it says **Emergency**. Tap there and then on the next screen, that same area in the lower left hand corner will now display **Medical ID**.

Tap it to see the information you typed in, as well as the **EMERGENCY CONTACTS** you specified. You can tap one of these contacts and your iPhone will dial the number, even though it is still locked.

Be sure to update this information every once in awhile and definitely help out a friend and tell them about this potentially life saving feature.

THIS IS another one of those iPhone features you hope you never have to use, but it's better to have it set up just in case.

Emergency SOS works in two ways. First, it can dial the proper emergency number for your location fast and automatically. I know you know the emergency number in the United States is 911, but what if you're in Italy, Japan or Costa Rica?

Second, it can send a text message to **Emergency Contacts** you specify with your location, which will repeat every 4 hours for 24 hours.

To set up the feature, go into **Settings** > **Emergency SOS**.

There are several options to consider. The first is **Call with Side Button**. By default, to activate Emergency SOS, you would press and keep holding down the side button and either volume button.

If you toggle the option to turn on **Call with Side Button**, you can also activate Emergency SOS by pressing the side button fast 5 times.

Keep in mind options for activating Emergency SOS might be different depending on your iPhone model. Your settings screen will show the button combinations that work for your phone.

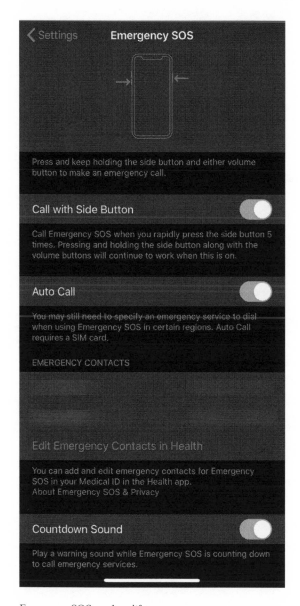

Emergency SOS can be a lifesaver

The second option is **Auto Call**. Toggle it on and your phone will automatically dial the proper emergency number for your location after a short countdown when you activate Emergency SOS using one of the methods described above.

You will have a short opportunity to cancel the call.

Toggle Auto Call off and when you activate Emergency SOS you will have to take an additional step to actually place the call. You'll see an Emergency SOS slider appear on screen - slide it to dial.

Auto Call on is easiest since there are no extra steps involved beyond activating the feature. However, if you think you might activate the feature by accident, it could be better to leave it off.

Personally, I don't want any extra time between me and emergency services.

The other thing to set up in Emergency SOS are your **EMERGENCY CONTACTS**.

You might already have some listed here if you set up **Medical ID**, if not, tap **Edit Emergency Contacts in Health** and you can choose the contacts that will be notified when you use the Emergency SOS function on iPhone.

They will also get your current location for a certain period of time if you ever activate the feature.

I'LL BET you've used the Control Center feature many times on your iPhone, but you've never called it that.

It's the area of the iPhone that you access by pulling down from the top right side of the home screen (or if you have an iPhone with a home button, you swipe up from the bottom of the screen to access it).

It's here where you can see options to put your phone into **Airplane** mode, toggle **Wifi**, change the brightness of the screen and more.

You can customize the buttons that appear here. Here's how.

Go to **Settings** > **Control Center** > **Customize Controls**. You probably already have some items added to Control Center, but you can pick and choose what you see.

Just press the **plus** or **minus** sign next to an item to add or remove it.

You might find it handy to have shortcuts to **Alarm**, an **Apple TV Remote** and **Calculator**.

Others might like to have **Dark Mode**, **Flashlight** and **Low Power mode** easily accessible with a tap.

Customize what appears in Control Center

Once you make your selections, you can change the order of your choices by using the "dragger" controls to the right of the items. They look like three little bars - just tap one to grab it and drag the item up or down.

You can get a look at your handy work by bringing up the Control Center after you've made some changes.

Keep in mind, certain controls can't be moved, including the radio toggles (for Airplane, WiFi, cellular and Bluetooth) and other system functions like screen rotate lock, Do Not Disturb and brightness.

You probably go into Control Center all the time to change things like the brightness and volume, but it might take some remembering to go here to perform the other tasks you just added.

Once you do, it can save you some valuable time normally spent searching for and adjusting these controls.

THIS IS the tip that started this entire book series, so let me show you a super easy way to access it.

Your iPhone can become a magnifying glass - helpful for seeing small text on say, a restaurant menu.

In my previous books, I showed you a way to activate the feature by turning it on and hitting the home or side button three times fast.

You can still do that by going to **Settings** > **Accessibility** > **Magnifier** > **On**.

But there's an even easier way to instantly turn your iPhone into a magnifying glass: add the button to **Control Center**.

Go into **Settings** > **Control Center** > **Customize Controls**.

Here, look for the control for **Magnifier** and tap the **plus sign** next to it.

Now, back out of this menu and go to your phone's home screen.

To turn your phone into a magnifying glass fast, access Control Center by pulling down from the top right hand corner of your screen (or swipe up from the bottom if you have a phone with a home button).

Add a Magnifier button to Control Center for fast access

You should see the same **Magnifier** icon. Tap it and your phone will bring up a magnifying viewfinder!

You can use the slider at the bottom to adjust the magnification level, turn on a light for illumination, snap a still to inspect something in more detail, or tap the filter icon to change the look of the text on the screen.

To exit the magnifier tool, just swipe up as you would from the bottom of the screen to exit any app or press the home button.

Reading a restaurant menu with the help of your phone will never be the same.

HOPE YOU'RE NOT in a comfortable seat, because once you read this tip you're going to want to get up and use it immediately.

It's how to turn on the built in level on your iPhone.

It's handy for straightening picture frames on the wall or making sure a table isn't lopsided.

To activate it, open the **Measure** app.

At the bottom of the screen, tap where you see the option for **Level**.

Immediately, your screen will start showing some numbers.

Hold your phone horizontal or vertical to see if something is level, or how many degrees it is off by.

When something is perfectly level, you'll feel a little vibration and the screen will light up green.

Lay your phone flat on a surface to see how level it is. In this mode, the numbers will appear in a circle, but again, when things are perfectly flat, your screen will light up green.

Remember the part where I said you'd want to use this right away? Go ahead, I totally understand if you want to take a break to run around and check all of your paintings, pictures and frames on the walls.

Just call me Rich on Interior Decorating.

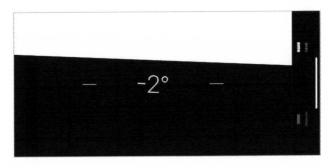

The Measure app has a Level tool inside

It's such a simple thing, but taking a screenshot on your iPhone can have so many uses.

You might want to remember a website you visited, grab a quick capture of something you want to buy or screenshot some text to stow away for inspiration later.

Whatever the reason, taking a screenshot is as easy as holding down two buttons simultaneously: the **Volume Up** + **Side Button**.

If your phone has a home button, press the **Power button** + **Home Button** at the same time.

Tap both fast at the same time and your phone will make a capture sound. Your screenshot will appear in the lower left hand corner of the screen.

It will go away after a moment or two, but it will be safely stored in your **Camera Roll**.

If you want to edit your screenshot or send it off to someone, just tap the thumbnail to bring up some helpful tools.

Apple has tweaked these editing tools in iOS 13, giving you even more options and flexibility when it comes to marking up your screenshot.

Screenshots are so last second (I know, really bad joke)

Let's check out some of the new features here.

For starters, look at the top row, above the screenshot. You'll see a new **Trash Can** button, which is helpful for deleting your screenshot instantly. Before, this action would take two steps - now it can be completed in just one.

The **share button** has moved to the upper right hand corner. You can tap here to access all of your share options for your screenshot.

If you look at the edges of your screenshot, you'll notice little **handles**. These are used to crop your screenshot. Just drag one or a few to slice away the edges of your screenshot. Drag them back out to get the cropped part back.

Below your image, you've got markup tools including a **pen**, **highlighter**, **pencil** and **eraser**. Tap to select one of them, tap again for more options including the ability to make the lines thicker and adjust the opacity of the line using the slider.

There's a new look for the **selection tool** next to the eraser. It used to look like a lasso, but now it looks more like a pencil with a mysterious top. You can use this to trace a lasso around something you've drawn on the screen. You can then tap and move that item around on screen.

Need to draw a straight line? Just use the **ruler tool**. Place it on the screenshot then adjust the angle to the exact degree you need. Now, choose a writing tool and trace along the side of the ruler and you'll get a perfectly straight line.

The **color selector** tool now shows you more colors when you first tap it. This will change the color of your writing instrument.

Next to it you'll see a button with a plus sign inside. This brings up even more tools including one for **Text**, adding a **Signature**, **Magnifier**, and a way to change the opacity of your screenshot.

Underneath, there are more options for drawing squares, circles, speech bubbles and arrows.

Yeah, you can have some fun with your screenshots.

When you're finished, just hit the **Done** button in the upper left hand corner to save your work to **Photos**, or **Files**, which is new a new option available in iOS 13.

Of course, you'll still free to use the **Share** button to share out your screenshot in a myriad of other ways.

It almost makes you wonder why they didn't name the song "Hit me with your best screenshot."

You know I can't help myself. What kind of humor did you expect from a total nerd?

THE IPHONE COMES with a default email signature, and I bet you know what it is:

Sent from my iPhone

Now, back in the day when it was novel to send an email from your phone, this was probably cool. But these days, this is almost a given.

It's time to make your signature more useful.

To change your default email signature, go to **Settings** > **Mail** and scroll all the way down to the bottom.

You'll see an option at the end for **Signature**, along with those "signature" words "Sent from my iPhone.

Tap here to change this to something you want your outgoing email messages to say at the bottom.

For business, I recommend including at least your name, email and a phone number.

For fun, you could go with an inspirational quote, something funny, serious or just delete it altogether.

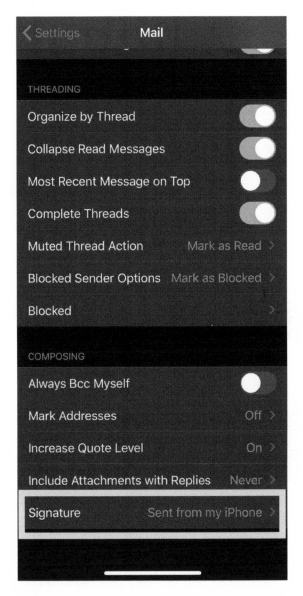

Change your email signature to something more original

IPHONE WIDGETS CAN BE a useful way to get helpful information at a glance. To see them, just swipe right to left or pull your home screen right.

Either way, this will reveal a page Apple calls the **Today View**. It's been around since the early days of the iPhone and comes pre-populated with several widgets including what's next on your calendar, Siri app suggestions and a few more things.

But you can make this screen way more useful just by taking a few minutes to decide what you want to see here.

To make it your own, tap the **Edit** button at the bottom of the **Today View** screen.

This will take you to a list of all of your widgets. The widgets listed in the top section are already on the page.

Those under **MORE WIDGETS** are available to add to the page.

Scroll through the list of available widgets and hit the **green plus sign** next to the ones you want to add to your page.

Alternatively, hit the **red minus button** next to those you don't need.

Tap Edit to choose your widgets

Since you can't see a preview of the information a widget will display, there might be some trial and error here. Add a few widgets, then hit **Done** in the upper right hand corner to go back to your Today View screen to see what the widgets show.

Then, hit **Edit** once again to go back to complete your customization. You might have to do this a few times, which is OK.

Sometimes I'll compare the information offered by two widgets - say Weather widgets - to see which one better suits my needs.

When you're happy with your widget selection, the last thing to do is to put them in an order you want.

From the **Add Widgets** screen, use the **handles** (they look like three lines) next to the name of each widget to drag it up or down in the list.

Hit **Done** when you're finished and you can now admire your handy work.

I use **Today View** for a quick glance at my next appointment, my Activity, some news headlines, sports scores, battery levels and more.

Don't forget, you can access this view from the lock screen just by swiping right or pulling left. Just remember, you might have to unlock your phone to see more sensitive data, like calendar appointments.

EVERY APP you install on your phone wants to send you notifications. These little messages can range from nagging to useful. Notifications are an app's livelihood: they remind us to interact with the app from time to time.

App developers know that most of us are too busy to deal with all of these notifications and we'll probably never turn them off.

But you're reading my book on the iPhone, so you're going to be smarter than that.

The next time a notification comes in, you can instantly manage it.

Instead of clearing it away, **swipe the notification right to left** to bring up a few options including a **Manage** button.

Tap here to bring up notification options that are otherwise hidden deep inside the Settings menu.

The two big buttons that appear here are the simplest options for notifications triage. **Deliver Quietly** means that the notification won't light up your screen or make a sound as it arrives. You also won't see a notification banner or a little red circle on the App itself.

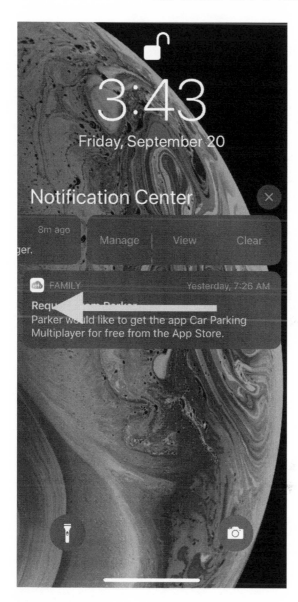

Swipe right to left on a single notification to manage its
settings

You will be able to see it if you check all of your notifications by
swiping up to see what I call the notification graveyard (see the next
tip).

This is handy for apps that aren't totally bugging you and you wouldn't mind seeing but you won't be interrupted by their notifications.

The next option is a bit more extreme: **Turn Off**... Tap this button and you won't see any notifications from that app at all.

Zero.

If you want to turn them back on, you'll have to go into **Settings** > **Notifications**, then find the app and turn on the options for **Sounds and Badges**.

CHANCES ARE, you get a lot of notifications on your phone.

Most of the time, they're arriving at various points throughout the day and you probably don't have a lot of time to deal with them.

Apple has realized that most people are terrible about clearing out their notifications, so they make it look like they've gone away even though they haven't.

Let me show you what I call the "notification graveyard."

First, just tap your phone so the lock screen lights up. You might see a recent notification or two on your lock screen.

Now, place your finger in the middle of the screen and swipe up a bit. Did you just drag up a bunch of older notifications that came to life on your screen? Welcome to the notification graveyard. It's where old notifications you haven't touched lay to rest.

You can go through these older notifications, one by one, swiping them away by dragging on them from right to left, then hitting the **Clear All** button, but that might take a while.

Here's a faster way.

See the **X** in a circle next to the words Notification Center?

This button can help you clear out both your old notifications, and optionally, your old and recent notifications.

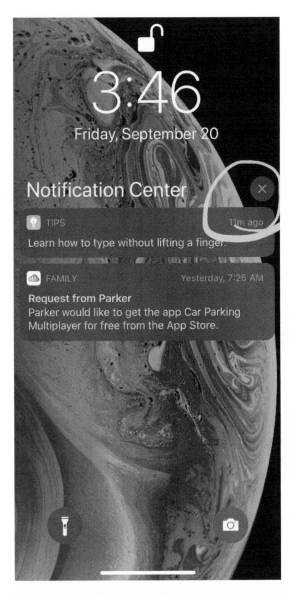

Press and hold this X to reveal a Clear All Notifications
button

To clear out the old notifications, just tap the **X** in the circle once.
It will switch to the word **Clear**. Tap it again to get rid of your old
notifications instantly.

You'll notice that your recent notifications still hang around.

Want to get rid of your old and recent notifications all at once? Instead of tapping the x in the circle, **tap and hold it** until you see a new, big button appear that says **Clear All Notifications**.

Give this a tap and - poof! - all of your notifications, recent and old, are cleared away instantly.

Doesn't it feel good to do that?

My favorite hidden button on the iPhone

HEY, Siri is an easy way to control your phone hands free, but once you turn the feature on in settings, it can feel like your phone is always listening to you.

Well, it is, but Siri is only listening for the phrase that pays, aka, the hotword or keywords "Hey, Siri."

But maybe you've used the iPhone long enough to know that Siri isn't always perfect, and sometimes she will respond even when you don't say those magical words.

Next thing you know, your iPhone screen is transcribing everything you say or you get an odd verbal reaction from your virtual assistant.

If you want to keep Siri from listening for a bit, all you have to do is flip your phone over.

That's right: when your iPhone is face down, listening for Hey, Siri is temporarily disabled. iPhone uses its fancy sensors to know when its screen is face down on a surface and will automatically pause listening for these two magic words.

Try it: first, ensure that listening for Hey, Siri is turned on by going into **Settings** > **Siri & Search** and making sure the toggle is on for **Listen for "Hey Siri**." If it's the first time you're turning it on, you will have to train Siri to recognize your voice.

Once that's complete, lock your phone and say the words, "**Hey Siri, what's the weather like outside?**"

Your phone should spring into action with an answer.

Now, flip your phone over so it's face down and ask the same question again.

There shouldn't be a response.

So, the next time you need some privacy, just flip your phone over and Siri will mind her own business.

It amazes me that even in iOS 13, there is still no way to have an alarm that isn't blaring in your ears at the crack of dawn.

Sure, you can change the volume of the alarm sound, but I like to ease myself awake.

Thankfully, there is a way to gently start your day, but it involves **Bedtime** instead of the standard **Alarm**.

To set it up, go into the **Clock** app. Now, normally, you're probably setting alarms with the Alarm tab at the bottom or just asking Siri to set your alarm for a certain time each morning.

But to get a gradual wakeup alarm sound, you'll have to use **Bedtime** at the bottom of the screen. Tap it and you might have to do some initial setup.

It will ask for your wake up time, your preferred alarm sound, bedtime and alarm schedule. Don't worry, you can always change these later.

Once you've set these items, you're good to go. Not only will your phone give you a notification before it's time to hit the sack, it will also automatically turn on **Do Not Disturb During Bedtime**, which darkens the lock screen and hides notifications while you're sleeping.

Wake up to a gradual alarm with Bedtime, but I don't recommend these times

No more opening one eye in the middle of the night, tapping your phone and being woken up by a bright screen full of messages you suddenly feel the urge to respond to.

But the best part: the alarm sound will start off soft and gently increase in volume until you wake up. It's quite a different way to start your day.

If you need to change your wake up time for a particular day, just go into Bedtime and tap the area where it says **Wake Up**. Use the slide wheel to adjust your **Bedtime** and **Wake Up** times.

Back on the main screen, you'll see a button labeled **Options** above **Bedtime**.

Tap it to tweak all of your Bedtime preferences, including when you get a reminder to go to sleep. You can also choose your **Wake Up Sound** and adjust the volume slider to a maximum volume that works for you.

Want a little preview of how your increasing alarm will sound? Just tap the circle on the volume slider to get an idea of what the gradual volume increase will sound like in the morning.

Sweet dreams!

WHY PAY to purchase or rent a TV show or Movie when it might be streaming for free?

There's a super simple way to check using the **TV** app on the iPhone.

To try it, just open the **TV** app and search for a movie or TV show in the search bar.

Once you find the item you're looking for, tap to select it.

On this screen, you'll see big blue buttons that will let you Buy or Rent the movie, but we're more interested in the less obvious button underneath that says **Open In**... or simply **Play** if you already have the app installed on your device.

Tap **Open In**... to see a list of all the streaming services that are showing that title for free.

If you see one that you subscribe to, you can watch the title there for free!

Search in the TV app to see if a movie or TV show is
streaming for free or included in your subscriptions

It might even be streaming on one of the free services like Pluto
TV, Tubi TV and others. Of course, there might be some ads sprin-

kled in on these totally free services, but you can make the final call whether to watch.

Apple is searching through a wide range of providers to come up with this information, and it's probably one of the best ways to see if something is streaming for free without hopping from service to service or checking a bunch of websites.

Nearly all of the major streaming providers are represented in the search, including Amazon Prime Video, HBO, Netflix and Showtime, as well as more niche sites including CuriosityStream and Shudder.

Look for the Open In... option for free streaming

You can see the full list using the link below. You'll find Apple's solution pretty comprehensive. This tip alone could save you the price of this book after just a few free movie nights.

https://support.apple.com/en-us/HT205321#usa

EVER MAKE a mistake on the iPhone calculator? You won't find a delete key to erase the last digit you pressed. There is a clear button, but that clears out everything.

Sometimes, you just want to erase a digit or two, and now you'll know the way to do it.

First, open the **Calculator** app. Then, type a few numbers.

Now we're going to erase one but the way to do it is actually hidden.

It's not a key but a swipe.

With some digits on the screen, just **swipe left or right in the blank area** right above the numbers. You'll notice that the last digit disappears.

Swipe again, the next digit disappears.

It doesn't matter if you swipe left or right, the result is the same. You just found the delete key on the iPhone calculator.

Congratulation. Your math problems will never be the same.

Swipe to erase a digit on the Calculator

99 / COPY AND PASTE FROM YOUR IPHONE TO ANOTHER APPLE DEVICE

It's no secret that Apple devices are tightly integrated. You might get your iMessages delivered to your iPhone and Mac computer, or save an article on your iPad and to read later on your iPhone.

This little trick will show you how to copy text from one device and paste it on another.

Apple calls it **Universal Clipboard**. To enable it, both of your devices must be signed into the same iCloud account, located near each other, and have Bluetooth turned on.

You'll also need to make sure a feature called Handoff is enabled on both devices.

On the iPhone and iPad, go into **Settings** > **General** > **Handoff** and ensure it's toggled on. On a Mac computer, go to **System Preferences** > **General** and make sure **Allow Handoff**... is enabled.

Now, to try this trick you'll need two devices. Let's open the Notes app on both devices and start up a fresh, different, note on both of them.

Type a word into the note on your first device: testing.

Next, highlight the word by tapping and holding on it to bring up a copy and paste menu. Tap **Copy**.

Activate Handoff to copy and paste between your Apple devices

On your second device, use the command for Paste. If it's a Mac, you can use the **Command + V** shortcut. If it's an iPhone or iPad, just tap the screen near the cursor and hit the **Paste** button that appears.

If all goes well, the word "Testing" should paste onto the second note!

Pretty neat, right? You can use this for web addresses, maps addresses and any other text you want to quickly copy and paste from one device to another.

Just keep in mind you have to copy and paste within a reasonable amount of time for security reasons. You can't copy today and paste tomorrow.

What kind of monster does that, anyway?

No MATTER how much iCloud space you've purchased, it always seems to be nagging you to get some more. I'm going to show you an overlooked place that might be hogging space without you even knowing it.

It usually happens when you upgrade to a new iPhone or iPad. Once an iCloud backup is used to move your data to the new device, the old backup remains, and it could be taking up gobs of valuable space.

Here's how to check to see if this is affecting your iCloud storage.

Go into **Settings** > **Apple ID, iCloud, iTunes & App Store** > **iCloud** > **Manage Storage**.

At the top of this screen, you'll see a breakdown on what is taking up your iCloud storage. If the section of the bar graph that represents Backups is a large portion, you might be in for a treat. Sure, it's in the form of reclaimed iCloud space, which is not very tasty, but still very exciting.

Tap the section labeled **Backups** and this will take you further into the depths of your mysterious iCloud storage.

Check to see if old Backups are taking up your iCloud storage

Here, you will see a list of all of your **BACKUPS**, which should be your current phone, along with any previous phones, iPads and other assorted Apple device you've owned over the years.

These backups can be quite large depending on what they contain inside them, which could be photos, documents, videos and whatever other data you have on your phone.

Tap a backup to get more information about it including the **Last Backup** date.

If you're worried about deleting the backup from the device you're currently using, you can tell which one it is because under the name it says "This iPhone."

If you find a backup you no longer need, you can hit the **Delete Backup** button at the bottom of the screen.

Just be sure before you delete a backup you truly no longer need the data inside.

Again, if you restored a phone from a backup, like when you upgraded your iPhone, the old backup is generally OK to get rid of.

It just doesn't happen automatically. Just make sure your data is safely migrated to the new device, and that device is backed up to iCloud.

Sometimes, the simplest tasks can prove tricky on an iPhone.

Like how to turn the thing completely off.

Ever since Apple ditched the power button, this seemingly simple task can have you spinning your head trying to figure out how to do it.

On the latest iPhones, it might seem like you would just press and hold down the side button until the device powers off.

But nope, that just activates Siri.

It actually takes two keys to turn off the iPhone. You have to **simultaneously press and hold down both the Volume Up Key + the Side key**.

After a few seconds of holding both down at the same time, you'll get a new screen with an option to "**slide to power off**."

Slide that slider and your phone will power off.

Turning it back on is much easier: just press and hold the side button until you see an Apple logo.

There is another way to shut down your iPhone: Settings.

Go into **Settings** > **General** and scroll all the way down until you see **Shut Down** in blue. Tap it to bring up the same "slide to power off" slider. You know what to do.

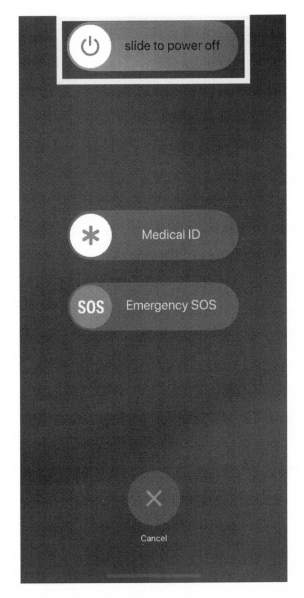

The elusive "slide to power off" switch

Cycling the power on your phone can also be handy if it's acting kind of wonky.

Although iPhones rarely need to be powered down, it can help if something seems to be acting up.

Think of it as a reset button for your device.

I hope you enjoyed learning more about the capabilities of your iPhone.

This is the third edition of my book and I've had a blast exploring all of these little features that make an iPhone so magical. I've tested dozens of phones over the years from various manufacturers and I've come to the conclusion that what makes an iPhone so special is that there is more than meets the eye.

At its heart, it's just a slab of metal and glass. Anyone can pick it up and make calls, take pictures and send messages. But dig a little deeper and it can become so many different things thanks to software, sensors and the camera.

From a measuring tape to magnifying glass, document scanner to audio recorder, Apple has managed to build in dozens of useful features. If something isn't included, a quick download of an App can introduce new functionality instantly.

It is my sincere hope that this book has at least opened your eyes to the potential of this transformational device. People may argue that we're spending too much time on our phones, and that's probably true, but when in history have we had one device do so many things, right in the palm of our hand?

I don't expect you to remember every tip I wrote, but at least you have an awareness of what your phone can do. The next time you need to scan a document or find an attachment buried deep inside your iMessages, you'll know where to find the answer.

ABOUT THE AUTHOR

Rich DeMuro is the tech reporter for KTLA-TV in Los Angeles and is known as "Rich on Tech" to viewers across the country. Rich is also a regular on KFI-AM 640 radio in Los Angeles, where he talks about the latest tech topics weekly.

In addition, Rich fills in for Leo Laporte on his nationally syndicated radio show "The Tech Guy." Rich also hosts a podcast called "Rich on Tech" where he talks about tech news and answers the questions people send him.

Rich has worked as a reporter in Washington, Louisiana and was senior editor at the technology website CNET in New York City.

Rich is originally from New Jersey and live in Los Angeles with his wife and two sons. He enjoys movies, running, magic and reading.

Sign up for Rich's newsletter at richontech.tv.

 facebook.com/richontech

 twitter.com/richdemuro

 instagram.com/richontech

Made in the USA
Coppell, TX
28 November 2019

11953658R00169